T0214311

Theory of Fuzzy Computation

International Federation for Systems Research International Series on Systems Science and Engineering

IFSR was established "to stimulate all activities associated with the scientific study of systems and to coordinate such activities at international level." The aim of this series is to stimulate publication of high-quality monographs and textbooks on various topics of systems science and engineering. This series complements the Federations other publications.

A Continuation Order Plan is available for this series. A continuation order will bring delivery of each new volume immediately upon publication. Volumes are billed only upon actual shipment. For further information please contact the publisher.

Volumes 1–6 were published by Pergamon Press

For further volumes:
http://www.springer.com/series/6104

Apostolos Syropoulos

Theory of Fuzzy Computation

 Springer

Apostolos Syropoulos
Xanthi
Greece

ISSN 1574-0463
ISBN 978-1-4939-4362-3   ISBN 978-1-4614-8379-3 (eBook)
DOI 10.1007/978-1-4614-8379-3
Springer New York Heidelberg Dordrecht London

Mathematics Subject Classification (2010): 68Q05, 68Q10, 68Q45, 68Q85.

Printed on acid-free paper

Springer is part of Springer Science+Business Media (www.springer.com)

To my son Demetrios-Georgios
and my father Georgios

Preface

Nowadays, computing is ubiquitous and pervasive. Many consumer products, like electronic and electric appliances and vehicles, are using some sort of computational device to ensure proper operation. At the same time, fuzzy set theory (roughly, the theory that claims that elements belong to sets to some degree) gains momentum as a basis for the solution of many problems in engineering, technology, medicine, etc. Thus, it seems that some sort of *fuzzy computation* would be a very important development. Indeed, since the inception of fuzzy set theory in the mid-1960s, there have been attempts to define fuzzy conceptual computing devices and so to develop a formal theory of fuzzy computation. These efforts yielded important results quite recently, and they include the discovery that there is no universal fuzzy Turing machine and that fuzzy Turing machines are capable to solve problems that no ordinary Turing machine can solve. Simultaneously, work has been carried out to design and implement fuzzy hardware (e.g., see [70] for a not so recent overview of the field of fuzzy hardware).

Unfortunately, even today many researchers and scholars confuse the notion of fuzzy computing with fuzzy expert systems, fuzzy database systems, fuzzy information management, fuzzy knowledge management, fuzzy e-commerce services, fuzzy web services, etc.—all these are "applications" of fuzzy set theory that are implemented atop normal hardware using ordinary computing tools. In a sense, fuzziness appears to be some sort of linguistic layer that facilitates the expression of certain problems, something completely unacceptable in my own opinion. In different words, there is a gap between what most people think fuzzy computing is about and what it should be about in fact. Now, fuzziness is not a linguistic phenomenon; it is a mathematical model of vagueness, which, in turn, is a characteristic property of this cosmos (Fig. 1 shows what I consider as vagueness in nature). For instance, one may argue that the probabilities employed in quantum mechanics are in fact *possibilities* in the sense of Zadeh's theory of possibilities [147] (e.g., see [40] for an early and brief discussion of this idea). Thus, there are a number of misconceptions that have to be clarified.

This book, among others, is an effort to remedy this deficiency. It presents most, if not all, milestones in the development of what one might call a theory of fuzzy computation. In particular, the first chapter starts by giving a historical overview of computation and fuzziness and gives a preliminary response to the question what is fuzzy computation,

Chapter 2 is a brief overview of the classical theory of computation. It includes a thorough presentation of Turing machines and some of their variations (e.g., multitape and nondeterministic Turing machines). Then, there is an introduction to Kolmogorov–Uspensky algorithms and a discussion of their computational power. This discussion is followed by a

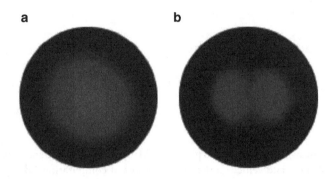

Figure 1: Vagueness in nature—Images of the end atoms of carbon chains. Reprinted figure with permission from I.M. Mikhailovskij, E.V. Sadanov, T.I. Mazilova, V.A. Ksenofontov, and O.A. Velicodnaja, PHYSICAL REVIEW B, 80, 165404, 2009. Copyright (2009) by the American Physical Society

presentation of recursive functions, the analytical hierarchy, and a discussion of the Church–Turing thesis.

The third chapter is an introduction to the theory of fuzzy sets. The chapter starts with some necessary definitions and results from order theory and proceeds with the definition of fuzzy sets and L-fuzzy sets. Next, there is a brief presentation of fuzzy relations, t-norm, and t-conorms. The chapter concludes with some thoughts concerning fuzzy set theory in general.

A thorough presentation of fuzzy Turing machines is given in chapter four. There is also a discussion of fuzzy formal languages, fuzzy recursion theory, fuzzy universality, and the computational power of fuzzy Turing machines.

The fifth chapter presents other models of computation that are inspired by fuzzy set theory. In particular, there is a discussion of fuzzy P systems, fuzzy labeled transition systems, fuzzy X machines, and the fuzzy chemical abstract machine. In addition, there is a discussion of some fuzzy process algebras.

Finally, the book has two appendices—one that discusses Zadeh's idea of *computing with words* and one that introduces rough computing devices. Strictly, computing with words is not a model of computation per se but an idea that is supposed to help people to solve problems where, for example, quantities are not expressed by numbers but by words like "much" or "too little" instead. Rough sets are an alternative approach to describe vagueness that makes use of an upper and a lower approximation. Since this book is about (conceptual) computing devices that are vague and operate in a vague environment, I felt it was more than necessary to include a short appendix on the emerging field of rough computing devices.

Intended Readership The book is self-contained, but the text assumes familiarity with certain mathematical notions (e.g., readers are expected to be familiar with relations and their properties). However, everyone with some elementary mathematical maturity willing to get a thorough understanding of the theory of fuzzy computation as it currently stands can read the book by skipping the difficult parts. Also, the book can also be used as a teaching vehicle for a graduate-level course in fuzzy computability theory. In fact, the author has followed the book's exposition for a short course on the theory of fuzzy computation that was taught at COPPE/UFRJ, the Alberto Luiz Coimbra Institute for Graduate Studies and Research in Engineering of the Federal University of Rio de Janeiro, in June 2012.

Acknowledgements   First of all I would like to thank Vaishali Damle, Springer's former Mathematics Senior Editor, for her help and assistance. I would like to thank Francisco Antônio de Moraes Accioli Dória for inviting me to Rio. Also, I would like to thank Barbara Beeton for her help. Last, but certainly not least, I would like to thank Dimos Mourvakis and Vassilis Papadopoulos for their constant support.

Xanthi, Greece   Apostolos Syropoulos

Contents

1. Introduction

The fusion of computability theory (CP) with fuzzy set theory (FST) demands a good knowledge of both theories. Thus, this introductory chapter tries to familiarize readers with a number of concepts and ideas that are necessary for the rest of this book. In particular, I start with a review of the events that lead to what is now known as computer science and then I give an overview of FST. The chapter concludes with a description of what the fusion of CP with FST is about and what to expect from it.

1.1 From Numbers to Computers

Since the Cradle of Humankind, numbers have played a very important role. They have influenced our culture and our language. Numbers have been used to *count* objects, animals, etc., and to *calculate* areas, quantities, volumes, etc. An ordinary person knows by heart a number of trivial arithmetic results and so she can perform some simple arithmetic operations in her head. Nevertheless, in the majority of cases, arithmetic calculations are complex or even very complex, so an ordinary person cannot perform them in her head. Thus, one has to use her fingers, paper and pencil, or some specially designed device that can assist her in the calculations. To the best of my knowledge, the first such assisting device is the abacus. Although one may argue that our fingers are in fact the first calculating assisting device, nevertheless, for reasons of simplicity I will not consider our fingers to be a calculating assisting device.

In the course of time, people had to perform more and more complex calculations and it was more than imperative to have someone or something whose sole task or job would be to rapidly and correctly perform calculations. Ingenious people, like Blaise Pascal, designed mechanical devices that would, almost, automatically perform additions, subtractions, etc. On the other hand, humans, which later were known as computers, were trained and, later on, employed to perform series of difficult and complex calculations. Both these human computers and the various mechanical devices had something in common—they were dully performing a number of steps to derive a particular result. Unfortunately, one serious drawback of these mechanical devices was that one could not instruct them to automatically perform different tasks. A solution to this problem was given by Joseph Marie Jacquard who was the first person who realized that one could instruct a machine to perform different tasks of the same kind. His mechanical looms were able to produce fabrics or cloths having different weaving. Following Jacquard's idea, Charles Babbage designed, but actually never completed the construction of his *analytical engine*, that is, a mechanical general-purpose computer. For reasons of completeness, let me say that Percy Ludgate designed also another analytical engine, which

A. Syropoulos, *Theory of Fuzzy Computation*, IFSR International Series on Systems Science and Engineering 31, DOI 10.1007/978-1-4614-8379-3_1,
© Springer Science+Business Media New York 2014

was never built also. Of course, one should not forget to mention Herman Hollerith who developed a mechanical tabulator that processed punched cards to rapidly tabulate statistical data. Finally, in the twentieth century, electronic digital computers were designed and built as a result of the demand for large-scale computing (see [56] for a historical account of this development).

Devices like the analytical engine seem to fundamentally differ from what an average person comes to understand as a computer. Generally speaking, computers, as we understand them today, use the binary numeral system to perform their tasks, while the binary numerals are encoded, in at least the early designs, by sequences of current pulses, whereas the mechanical computers used gears and other similar means to perform their tasks. Despite their *obvious* differences, all these machines have some common design principles since they operate in a discrete way, are programmable, and have the same computational power. But how do we know that they do have the same computational power? Moreover, how can we compare the computational power of two different computing devices? In order to answer these and similar questions, we need to know about the basic ideas behind computing devices, in particular, and computation, in general.

Nowadays, it is widely accepted that the Turing machine, that is, the conceptual computing machine devised by Alan Mathison Turing, is the archetypal computing device, which *dictates* what can and what cannot be computed. Thus, one can compare the computational power of any device to the computational power of the Turing machine in order to determine the computational power of her machine. Interestingly, it is not universally accepted that the computational power of the Turing machine poses an upper bound to what can be achieved by any computational device (see [126] for a thorough discussion of the limits of computation). Thus, one could say that the Turing machine is a measure of the computational power of any computing device. Even so, it is quite interesting to know why the Turing machine played and still plays such an important role in the development of computability theory.

During the *Second International Congress of Mathematicians*, which was held in Paris in 1900, David Hilbert, the famous German mathematician, was invited to deliver one of the main lectures. Hilbert presented 23 major mathematical problems,[1] which were designed to serve as examples for the kinds of problems whose solutions would lead to the furthering of disciplines in mathematics. In particular, Hilbert's second problem is about the consistency of the axioms of arithmetic, that is, it asks whether the axioms are independent and, more importantly, not contradictory. In his own (translated) words [65]:

> Upon closer consideration the question arise: Whether, in any way, certain statements of single axioms depend upon one another, and whether the axioms may not therefore contain certain parts in common, which must be isolated if one wishes to arrive at a system of axioms that shall be altogether independent of one another.
>
> But above all I wish to designate the following as the most important among the numerous questions which can be asked with regard to the axioms: To prove that they are not contradictory, that is, that a finite number of logical steps based upon them can never lead to contradictory results.

1. As noted in [41, p. 377], Hilbert actually presented ten of the problems to his audience. All 23 problems are listed in the printed form of his lecture.

What is not clear from this excerpt and the previous discussion is that Hilbert envisioned to transform mathematics into a *formal system* or *calculus*. Many fundamental parts of mathematics, including arithmetic, are *deductive systems*. Any such system includes a limited number of axioms, whereas each theorem follows logically from the axioms and/or any theorems, which might have been shown earlier, according to a limited number of inference rules. By representing axioms and theorems by *formulas*, that is, sequences of arbitrary symbols (i.e., strings), and the inference rules by *transformation rules*, which specify how a string can be converted into a new string, one ends up with a symbolic representation of a deductive system. Any such symbolic representation is a formal system. A complete sequence of symbolic manipulations in the calculus corresponds to and represents a sequence of deductions in the deductive system. If we reverse this process, that is, we first construct a calculus and then find an interpretation of it so it represents a deductive system, we arrive at the idea behind Hilbert's *metamathematics*. In different words, what Hilbert was seeking was to transform mathematics to a formal activity that would solve all possible problems. Obviously, this formal activity is reminiscent of programming as a mental activity. Now, what Hilbert's second problem was asking is whether it is possible to decide if any one axiom of a calculus is derivable in part or whole from the other axioms of the calculus. In addition, he was asking if calculi are *consistent*, that is, if it is impossible to derive from the axioms of a calculus two contradictory formulas (i.e., formulas like $3 < 2$ and $3 > 2$).

At the 1928 International Congress of Mathematicians in Bologna, Hilbert introduced a new problem—the so-called Entscheidungsproblem, that is, a problem that can be answered with *yes* or *no* (i.e., a decision problem). In particular, Hilbert asked whether or not it is possible to prove every true mathematical statement (see [133] more details). It is not an exaggeration to say that Hilbert was dreaming of a completely mechanized mathematical science where theorems could be proved automatically by some machine. More precisely, Hilbert asked whether calculi are *finitely describable* (i.e., whether the axioms and transformation rules are constructible in a finite number of steps, while, also, theorems should be provable in a finite number of steps), *complete* (i.e., whether every true statement that can be expressed in a given calculus is formally deducible from the axioms of the system), and consistent.

Three years after Hilbert's Bologna address, a young Austrian mathematician named Kurt Friedrich Gödel published one of the most important papers of the twentieth century that turned Hilbert's fondest dream into a nightmare. Gödel [53, 96], after defining a calculus that represents (a form of) arithmetic, proved that there are propositions (i.e., formulas of these calculus) that are *undecidable*, that is, one cannot decide if they are provable or not, within his calculus. In addition, he found a proposition which together with its negation cannot be obtained from the axioms and applications of the transformation rules. Thus, he managed to find a calculus that is incomplete.

A few years later, Alonzo Church [25, 26] introduced his λ-calculus,[2] which was intended to serve as a foundation for mathematics. Roughly, in the λ-calculus one defines functions of one argument using the notation $\lambda x.E$, where x is a bounded variable that may occur one or more times in the expression E. If E is a function and F a value, then the expression EF denotes the application of E to F. The result of the application is obtained

---

2. Nowadays, the term λ-calculus refers to a calculus that is used as a means of describing purely syntactically, the properties of mathematical functions, effectively treating them as rules.

by the syntactic replacement of each occurrence of the bounded variable by F. Although Stephen Cole Kleene and John Barkley Rosser [74] proved that the full system is not consistent (strictly speaking it is not ω-consistent), Church [28] showed that a consistent subsystem, dealing with functions only, was incomplete. In particular, he showed that given two formulas, it is not possible to say whether the first is *convertible* into the second. Since convertible may mean, among other things, identical, one cannot decide if two functions are equal. In addition, he identified the notion of an *effective calculable*, or effectively computable in modern parlance, function of positive integers with the notion of a λ-definable function of positive integers. Church [27] went one step further and showed that the Entscheidungsproblem is unsolvable.

At about the same time, Turing [135] introduced a conceptual machine that now bears his name. Roughly, this machine consists of a tape that is divided into editable cells whereas a scanning head can read the contents of cells. At any given moment the machine is in some state. Depending on the current state and the contents of the cell that the scanning head reads, the machine either replaces the contents of the cell or moves to the right or to the left. In all cases, it changes its current state. All actions of the machine are determined by consulting a set of instructions. Since each Turing machine is designed to solve a particular problem, Turing devised a *universal* machine that can emulate the behavior of any other conventional machine by employing a technique invented by Gödel. Loosely speaking, Gödel devised a procedure by which one can transform any string that belongs to a formal system, or formal language in modern parlance, into a unique natural number. Thus, Turing showed how to transform a Turing machine and its input into unique natural numbers that are processed by his universal machine. Also, he showed that this universal machine cannot determine whether a given machine that runs on specific input, which is printed on its tape, will terminate or not. This problem is known in the literature as the *halting problem*. Turing then used this result to show that the Entscheidungsproblem is unsolvable. In particular, he showed that there is no general method to tell whether a given formula is provable in some specific formal system. Furthermore, he showed that every λ-definable function of positive integers is *computable* (i.e., it can be calculated by his machine). In different words, he defined that to effectively calculate a function value means that one can define and, consequently, "use" a Turing machine to compute the same function value.

Loosely speaking, an *algorithm* is a "clerical (i.e., deterministic, bookkeeping) procedure which can be applied to any of a certain class of symbolic *inputs* and which will eventually yield, for each such input, a corresponding symbolic *output*" [109, p. 1]. Obviously, this is an informal "definition," albeit the notion of an algorithm is informal too. Despite of this, many identify algorithms with effectively calculable procedures. So in order to avoid confusion, I will say no more about algorithms unless it is absolutely necessary.

Turing's work had a great impact in what we call computer science today. In the early days of computer science, Marvin Minsky [93, p. 104] gave his understanding of the significance of Turing's work:

> This paper [i.e., [135]] is significant not only for the mathematical theory which concerns us here, but also because it contains, in essence, the invention of the modern computer and some of the programming techniques that accompanied it. While it is often said that the 1936 paper did not really much affect the

practical development of the computer, I could not agree to this in advance of a careful study of the intellectual history of the matter.

It is true that the first computers operated in a manner (superficially?) similar to the way the Turing machine operates; nevertheless, modern computers interact with their environment. But the elegance and the power of the Turing machine make some authors to forget this difference:

> He will unveil a Universal Computing Machine—today commonly termed the Universal Turing Machine or UTM—that is functional (if not exactly commercial) equivalent of a modern computer. [105]

Thus, it is important to emphasize that modern computers are not instantiations of Turing machines! Indeed, Jean-Yves Girard [52, pp. 409–410] in his balanced description of computer science takes this remark under serious consideration:

> Although man-made, computer science is the physics of logicians. In the beginning:
>
> - Many computer scientists failed to realize the infeasibility of the halting problem... with, as a result, the building of various paralogics.
> - More educated people were still paying much attention to formal issues, with an excessive, and surrealistic, emphasis on consistency, and the building of too many Broccoli logics.
>
> In general, computers prompted a renewal of positivistic nonsense, artificial intelligence and so on. But how unfair it would be to reduce computer science to these archaic mistakes. It is an immense source of intuitions, such as non-determinism, locations, proof-search, streams, process algebras... not to mention the mere idea of interactivity. Without computer science, would there still be any room left for logic?

In a nutshell, the Turing machine is the first realistic model of computation, which, at the same time, is the basis of modern computability theory, or recursion theory, as it is also known. Thus, anyone attempting serious work in the field of computability theory needs a basic understanding of Turing machines and recursion theory.

1.2 What Is Fuzziness?

Fuzziness is a mathematical model of *vagueness*, but, unfortunately, these terms are used (almost) interchangeably nowadays. It is widely accepted that a term is vague to the extent that it has borderline cases, that is, a case in which it seems impossible either to apply or not to apply a vague term. The sorites paradox (σόφισμα τοῦ σωρείτη), which was introduced by Eubulides of Miletus (Εὐβουλίδης ὁ Μιλήσιος), is a typical example of an argument that shows what it is meant by borderline cases. Also, the paradox is one of the so-called little-by-little arguments [68]. The term "σωρείτες" (sorites) derives from the Greek word σωρός (soros), which means "heap." The paradox is about the number of grains of wheat that makes a heap. All agree that a single grain of wheat does not comprise a heap. The same applies for two grains of wheat as they do not comprise a heap, etc. However, there is a point where the number of grains becomes large enough to be called a heap, but there is no general agreement as to where this occurs.

Bertrand Russell[3] [110] was perhaps the first thinker who has given a definition of vagueness:

> *Per contra*, a representation is *vague* when the relation of the representing system to the represented system is not one-one, but one-many.

For instance, Russel suggests that a photograph which is so smudged that it might equally represent Brown or Jones or Robinson is vague. Building on Russell's ideas, Max Black [17] had argued that most scientific theories, which I believe should include a theory of computation, are "ostensibly expressed in terms of objects never encountered in experience." In different words, one could argue that the Turing machine is an idealization of some real-world system and as such does not correspond to anything real! Black [17] proposes as a definition of vagueness the one given by Charles Sanders Peirce:

> A proposition is vague when there are possible states of things concerning which it is intrinsically uncertain whether, had they been contemplated by the speaker, he would have regarded them as excluded or allowed by the proposition. By intrinsically uncertain we mean not uncertain in consequence of any ignorance of the interpreter, but because the speaker's habits of language were indeterminate.

According to Black, the word *chair* demonstrates the suitability of this definition. But it is the "variety of applications to objects differing in size, shape and material" that "should not be confused with the vagueness of the word." In different words, vagueness should not be confused with *generality*. In addition, vagueness should not be confused with *ambiguity*. A term or phrase is ambiguous if it has at least two specific meanings that make sense in context. For example, consider the phrase *he ate the cookies on the couch*. Obviously, one can say that this means that someone ate his cookies on the couch or that someone ate the cookies that were on the couch. In conclusion, vagueness, ambiguity, and generality are entirely different notions.

It is widely accepted that there are three different expressions of vagueness [121]:

Many-Valued logics and fuzziness Borderline statements are assigned truth values that are between absolute truth and absolute falsehood.[4]

Supervaluationism The idea that borderline statements lack a truth value.

Contextualism The truth value of a proposition depends on its context (i.e., a person may be tall relative to American men but short relative to NBA players).

Since this book is about fuzzy models of computation, I will say no more about supervaluationism and contextualism. However, I believe it makes sense to say a few things about many-valued logics since, in a sense, they are forerunners of FST.

---

3. His full name was Bertrand Arthur William Russell, third Earl Russell.
4. Although Ivo Düntsch [44] has proposed a logic of rough sets (see Appendix B), still the field is in its infancy.

It seems that Aristotle (Ἀριστοτέλης) was the first thinker who recognized that there are propositions that cannot be classified as either true or false. In particular, in Chapter IX of his treatise *De Interpretatione* (Περὶ Ἑρμηνείας, On Interpretation), which is part of his *Organon* (Ὄργανον), he ponders about *future contingents* and their truth values. He concludes that [3]:

ὥστε δῆλον ὅτι οὐκ ἀνάγκη πάσης καταφάσεως καὶ ἀποφάσεως τῶν ἀντικειμένων τὴν μὲν ἀληθῆ τὴν δὲν ψευδῆ εἶναι· οὐ γὰρ ὥσπερ ἐπὶ τῶν ὄντων οὕτως ἔχει καὶ ἐπὶ τῶν μὴ ὄντων, δυνατῶν δὲ εἶναι ἢ μὴ εἶναι, ἀλλ' ὥσπερ εἴρηται.[5]

Jules Vuillemin [139] considers that Chapter IX of De Interpretatione is Aristotle's response to the *Master Argument* that has been recorded by Epictetus (Ἐπίκτητος). Interestingly, Vuillemin assumes that Aristotle's work can be correctly interpreted by using the notion of probability! In a sense, Aristotle's work influenced the development of many-valued logics by Jan Łukasiewicz and Emil Leon Post [57]. In a few words, a many-valued logic is a logic where propositions may assume more than two truth values. For example, a typical three-valued logic may have true (T), false (F), and undecided (U) as truth values. The following is the truth table for the logical conjunction of a three-valued logic:

$$T \wedge T = T \qquad T \wedge U = U \qquad T \wedge F = F,$$
$$U \wedge T = U \qquad U \wedge U = U \qquad U \wedge F = F,$$
$$F \wedge T = F \qquad F \wedge U = F \qquad F \wedge F = F,$$

One can define logics with a finite number n of truth values. It is convenient to represent these n truth values as fractions:

$$\frac{0}{n-1}, \frac{1}{n-1}, \dots, \frac{n-2}{n-1}, \frac{n-1}{n-1}.$$

It is possible to even define an infinite-valued logic where the truth values are elements of the set $\mathbb{N}$ of natural numbers including zero. A special form of infinite-valued logic is fuzzy logic. However, this logic was a "by-product" of FST, which will be discussed in the rest of this section.

Sets occupy a key position in mathematics. Indeed, one approach to the foundations of mathematics is based on the idea that sets are the most fundamental objects of mathematics. Roughly, a set is any collection, group, or conglomerate of things of any kind that are called *elements*. According to the established view, given an element x and a set A, then x either belongs to the set, denoted by $x \in A$, or does not belong to the set, denoted by $x \notin A$. By relaxing this requirement, one gets generalizations of the concept of a set. In different words, by generalizing the membership relation, one gets a generalization of sets. For example, if we allow multiple copies of an element to be members of a set, we get *multisets* [123]. Another way to generalize sets is to allow elements to belong to a set to some degree, which is the idea behind *fuzzy sets*, which have been introduced by Lotfi Asker Zadeh[144].[6] One should be aware that it is not enough to propose the generalization of some structure—one needs to

---

5. Translation: *It is clear then that it is not necessary for every affirmation or negation taken from among opposite propositions that the one be true, the other false. For what is non-existent but has the potentiality of being or not being does not behave after the fashion of what is existent, but in the manner just explained.*
6. While working independently from Zadeh, Dieter Klaua [72] discovered his *many-valued sets*. Interestingly, these sets are in a sense equivalent to Zadeh's fuzzy sets (see [60] for an overview of Klaua's work).

justify the generalization and, then, to show how new mathematics can be built from this generalization. Thus, in the rest of this section, I will try to address these matters.

Fuzzy sets have been introduced to model, among other things, cases where an object has some property to some degree. In the simplest case, this degree is a real number that belongs to the unit interval (i.e., the set $[0, 1]$), that is, if r is a membership degree, then $0 \leq r \leq 1$. Typically, there is a universe, that is, an ordinary set, which is used to define its fuzzy *subsets*.[7] For example, if we have a set of people, then we can define the fuzzy subset of the tall people. However, one should be aware that the assignment of membership degrees is not an algorithmic process. For instance, if Serena is 1.75 m tall and her friends are basketball players, which are considered tall people, and we are asked to form a fuzzy set of tall people, clearly her membership degree will be rather low. On the other hand, if she spends her free time with *average* people, then her corresponding membership degree will be probably higher. Interestingly, even if the membership degree of some element is zero, still this element belongs to the fuzzy subset.

Before presenting an interesting objection to the very idea behind fuzzy sets, let me stress that there are a number of scholars who believe that vagueness should be equated with lack of information. However, "[e]xperience has shown that no measurement, however carefully made, can be completely free of uncertainties" [132, p. 3]. For example, this is something I have learned during my first days at the university. At that time, I and my teammates had to perform our first laboratory exercise, which involved the calculation of the volume of a wooden cube. We were surprised by the *simplicity* of the assignment, yet we were also puzzled by the instructor's demand to measure the edge of the cube ten times. When we proceeded with the actual measurements, we realized that not all of them were the same! And if one cannot measure exactly the edge of a cube, then how can one be sure about the height of a person? Let me continue with the objection against FST.

A reader who has not bought the idea of fuzzy sets may think that instead of saying that Serena is tall to a degree of 0.70, we would state that *Serena is* 70% *tall* and, thus, we would have transformed a vague statement into an *exact* one! Indeed, it has been advocated that the need for fuzzy sets is just an "elementary mistake of logic" [24] (in the original example the author was arguing about cups and the degree to which they are full, but I believe this example serves the current discussion better). To begin with, the statement *Serena is* 70% *tall* can be either true or false. On the other hand, the statement *Serena is tall to a degree of* 0.70 is actually a "compound" statement that consists of the proposition *Serena is tall* and its denotation that happens to be the number 0.70 (1 denotes absolute truth while 0 denotes falseness). Nevertheless, one may argue that even the statement *Serena is* 70% *tall* is a "compound" one. However, "...the referent of probabilistic statements, in particular probability distributions, is the *abstract object* with which the theory of probabilities deals. Probability statements do not refer to individual objects which have the properties (among others) of this

---

7. Strictly speaking when defining a set using its *characteristic* function we do something similar (see Definition 2.4.3 on p. 34).

abstract object." [35, pp. 51–52]. Therefore, the name Serena in the probabilistic statement refers to some abstract Serena while the fuzzy statement to a specific one.

The great Canadian physician William Osler stated that "Medicine is a science of uncertainty and an art of probability." Statements like this are based on the traditional view that probability theory is the most important, if not the only, facet of uncertainty (e.g., see [81] for an example of a paper that defends this view). Although it has been argued that fuzziness is another facet of *uncertainty*,[8] and I will say more on this in a moment, still today there are researchers that work (?) in the area of FST and assume that fuzziness and probability theory are the same! Unfortunately, I have learned this the hard way. A couple of years ago, I submitted a paper to a prestigious journal devoted to the international advancement of the theory and application of fuzzy sets. The paper described a model of computation built on the idea that fuzziness is a fundamental property of our world. To my surprise, the paper was rejected mainly because three (!) reviewers insisted that it is a "fact that there is an equivalence between FST and probability theory." Obviously, I was shocked with what the reviews stated, but I thought that if reviewers and editors of journals, who are supposed to be devoted to the advancement of FST, express and endorse, respectively, such ideas, then people who are not convinced that FST has anything to offer should dismiss FST as almost nonsense! Of course, just because some people have not clarified their thoughts, to say the least, it does not mean FST is nonsense. On the contrary, the advent of FST has brought a new way of looking at the world that surrounds us.

In a response to a typical polemic against FST [81], Zadeh [148] advocated the idea that FST and probability theory are complementary theories and not competitive. In particular, Zadeh presented a number of reasons why probability theory cannot be used to tackle all problems that are encountered in an environment of uncertainty. Some of these reasons are presented briefly below:

(i) One cannot express in the language of probability theory judgments like *tomorrow will be a warm day* or *there will be a strong earthquake in the near future*. According to Zadeh these propositions involve *fuzzy events*.

(ii) One cannot express certain fuzzy quantifiers like *many*, *most*, and *few* in the language of probability theory.

(iii) It is not possible to perform estimations that involve fuzzy probabilities expressed as *likely, unlikely, not very likely*, etc.

(iv) It is difficult to analyze problems in which data are described in fuzzy terms.

FST gave birth to the so-called fuzzy logic, that is, a logic where propositions usually assume values in the unit interval (i.e., the set $[0, 1]$). In particular, a truth value equal to zero means that a proposition is false, while a truth value equal to one means that a proposition is (absolutely) true. One could say that fuzzy logic is an infinite-valued logic, since the unit interval has an infinite number of elements. Although fuzzy logic deals with fuzzy data, fuzzy terms, etc., still it is not fuzzy (enough?)! On the contrary, Zadeh [151] has argued

---

8. It is a fact that probability theory is tied to uncertainty, which can be associated with ignorance, while fuzzy sets are associated with vagueness, which has nothing to do with ignorance. Nevertheless, many researchers and thinkers tend to confuse the two terms.

that "fuzzy logic is a precise logic of imprecision and approximate reasoning." Furthermore, Zadeh [151] goes on to argue that

> [F]uzzy logic may be viewed as an attempt at formalizations/mechanization of two remarkable human capabilities. First, the capability to converse, reason and make rational decisions in an environment of imprecision, uncertainty, incompleteness of information, conflicting information, partiality of truth and partiality of possibility—in short, in an environment of imperfect information. And second, the capability to perform a wide variety of physical and mental tasks without any measurements and any computations.

Bart Kosko [80], another prominent fuzzy set theorist, defended the superiority of FST when compared to probability theory by saying that fuzziness "measures the degree to which an event occurs, not whether it occurs. Randomness describes the uncertainty of event occurrence." Although it is possible to find many more arguments that are in favor or against FST, most of them are not using results from physics. However, there is a recent development in quantum mechanics which I believe will have some impact in the battle between fuzzy set theorists and probability theorists. But before presenting this development, I will very briefly review the relevant ideas.

A number of scientific disciplines assume that everything that happens is the inevitable result of a cause that preceded in time. In different words, it is assumed that there is a predetermined course of events, similar to the notion of kismet. This view is called *determinism*. The opposite view of determinism is the doctrine of *free will* that assumes that when there are multiple possibilities, a person freely decides which to choose. Also, *combatibilism* is the view that determinism and free will are compatible. According to this view, our actions are partially free and partially deterministic, that is, our previous actions and prejudices may affect our *free* future actions (see [8, 118] for two excellent discussions of free will, determinism, and combatibilism).

Recently, John Horton Conway and Simon Kochen [33] have proved their *Free Will Theorem* (FWT), which roughly states that [33, p. 1444]:

> [I]f experimenters have a certain property, then spin 1 particles have exactly the same property. Since this property for experimenters is an instance of what is usually called "free will" we find it appropriate to use the same term also for particles.

Furthermore, they make the following interesting remark [33, p. 1465]:

> It is true that particles respond in a stochastic way. But this stochasticity of response cannot be explained by putting a stochastic element into any reduction mechanism that determines their behavior, because this behavior is not in fact determined by any information in their past light cones. This includes injected stochastic information that can be non-locally correlated.

In addition, Conway and Kochen [34, p. 209] conclude that

> '[a]lthough the FWT suggests to us that determinism is not a viable option, it nevertheless enables us to agree with Einstein that "God does not play dice

with the Universe." In the present state of knowledge, it is certainly beyond our capabilities to understand the connection between the free decisions of particles and humans, but the free will of neither of these is accounted for by mere randomness.

In different words, based on their work, Conway and Kochen rule out determinism as a view to explain the world, while, at the same time, they do not believe randomness plays the role it was assumed to play (indeed, their proof does not make any use of randomness). Obviously, Conway and Kochen have not considered using the theory of fuzzy sets to explain some aspects of the behavior of particles. In different words, it seems that vagueness and *nondeterminism* are primitive notions and as such they have a fundamental role to play.

Because the notion of a fuzzy set is an ingenious idea, we should have witnessed the development of truly fuzzy mathematics. Unfortunately, this is not the case and here is what Saunders Mac Lane had to say about this matter [90, pp. 439–440] :

> Not all outside influences are really fruitful. For example, one engineer came up with the notion of a *fuzzy set*—a set X where a statement $x \in X$ of membership may be neither true nor false but lies somewhere in between, say between 0 and 1. It was hoped that this ingenious notion would lead to all sorts of fruitful applications, to fuzzy automata, fuzzy decision theory and elsewhere. However, as yet most of the intended applications turn out to be just extensive exercises, not actually applicable; there has been a spate of such exercises. After all, if all Mathematics can be built up from sets, then a whole lot of variant (or, should we say, deviant) Mathematics can be built by fuzzifying these sets.

Of course, Mac Lane is wrong when he says that a membership degree lies between 0 and 1, since membership degrees can assume any values in the unit interval, but, obviously, this is not the point made in this statement. Despite some serious efforts to axiomatize fuzzy logic (e.g., see [58, 59, 62] for an overview of efforts in the right direction, although [59] misses some work done by this author [98, 99, 125]), to the best of my knowledge, no one has proposed a foundation of mathematics based on fuzzy sets. Also, very few mainstream fuzzy theorists have addressed some serious drawbacks of FST (e.g., see Sect. 3.6 for a brief discussion of why FST is not fuzzy enough). Clearly, this is not a goal of this book to tackle these problems, but it is my goal to show that fuzziness has something to offer in computability theory and that it is not just another extensive exercise.

1.3   Vague Computing in an Uncertain Environment

As is generally understood, computing is an exact activity. In different words, either a function is computable by some real or conceptual machine or it is not computable. However, it has been recognized that even partial results are useful (see [69]), which implies that even the partial computation of certain function values is of some use. The notion of a result to a certain degree, which is not the same as a partial result, computed by some real or conceptual fuzzy computing machine is also useful. Assume that someone has built a truly fuzzy computing device. Then, what do we expect to do with these machines? In other words, do we expect to be able to solve noncomputable problems with fuzzy computing machines? In what follows, I will try to answer this questions.

We know, due to Turing's work on the halting problem, that when a program gets stalled, we cannot say whether it has entered an infinite loop or not. In different words, one cannot tell, using "logical" arguments, whether a stalling program will definitely terminate or not. Not so surprisingly, any computing device, either real or conceptual, has its own halting problem, which, technically speaking, is not decidable by the machine itself. However, if a machine cannot solve its own halting problem, then there is no reason why another entirely different computing machine cannot solve the halting problem of the first. Thus, it makes sense to try to see whether fuzzy computing devices are able to solve the *classical* halting problem, at least up to some degree.

Consider the following problem: Given a finite set, Γ, of first-order logic sentences and a sentence D, is it possible to devise an effective method that will tell us in a finite amount of time whether Γ implies D? This problem is known as the decidability problem of first-order logic. It is known that if such an effective method exists, then the halting problem for Turing machines is decidable. Now, Petr Hájek [62] sees fuzzy logics as many-valued logics sui generis. Based on this, he has shown that certain fuzzy logics are undecidable, in the sense just explained, because some sets are not *recursive* (roughly, a set is recursive if one can devise an effective method that will tell us in a finite amount of time whether or not a given element belongs to the set). Of course the whole proof is irrelevant to our discussion but a skeptical reader may wonder whether this result is of any use. For example, can one use it to deduce that the halting problem of a fuzzy version of the Turing machine is undecidable? In order to be able to answer this question one needs to have a formal definition of the fuzzy version of the Turing machine in a fuzzy logic setting, then to show the existence of a universal fuzzy Turing machine, and, finally, to investigate the properties of the halting problem of this machine. Obviously, this implies that at the same time one needs to define the notion of a decidable fuzzy set. In other words, one needs to construct a new theory of fuzzy computation, which will possibly extend *classical* computability theory! To put it simply, it is necessary to have a fully developed theory of fuzzy computation in order to give definitive answers to questions like these.

Loosely speaking, a quantum computer is a computing device that directly uses quantum mechanical phenomena to perform operations on data. By analogy, one should expect that fuzzy computing devices should directly use fuzzy "phenomena" to perform operations on either fuzzy or exact, or *crisp*, data.[9] For instance, it makes sense to expect some fuzzy variants of the Turing machine to be able to determine which symbol is printed on a given cell with some plausibility degree. Also, it is quite possible that the same machine will perform some faulty printing, etc. Interestingly, Jarosław Pykacz, Bart D'Hooghe, and Roman R. Zapatrin [107] have shown that there is some connection between quantum and fuzzy computing.

---

9. In fuzzy theoretic literature, the term crisp is used to denote something that is exact.

2. A Précis of Classical Computability Theory

Turing machines form the core of computability theory, or recursion theory as it is also known. This chapter introduces basic notions and results and readers already familiar with them can safely skip it. The exposition is based on standard references [18, 39, 67, 83, 109]. In the discussion that follows, the symbol $\mathbb{N}$ will stand for the set of positive integer numbers including zero and $\mathbb{Q}$ will stand for the set of rational numbers.

2.1 Turing Machines

As explained in the introduction, a Turing machine is a conceptual computing device consisting of an *infinite tape*, a *controlling device*, and a *scanning head* (see Fig. 2.1). The tape is divided into an infinite number of cells. The scanning head can read and write symbols in each cell. The symbols are elements of a set $\Sigma = \{S_1, ..., S_n\}$, $n \geq 1$, which is called the *alphabet*. Usually, there is an additional symbol, $\sqcup$, called the *blank* symbol, and when this symbol is written on a cell by the scanning head, the effect of this operation is the erasure of the symbol that was printed on this particular cell. At any moment, the machine is in a *state* q_i, which is a member of a finite set $Q = \{q_0, q_1, ..., q_r\}$, $r \geq 0$. The controlling device is actually a lookup table that is used to determine what the machine has to do next at any given moment. In particular, the action a machine has to take depends on its current state and the symbol that is printed on the cell the scanning head has just finished scanning. If no action has been specified for a particular combination of state and symbol, the machine halts. Usually, the control device is specified by a finite set of *quadruples*, which are special cases of expressions.

Definition 2.1.1 An *expression* is a string of symbols chosen from the list $q_0, q_1, ...; \sqcup, S_1, ...;$ R, L.

A quadruple can have one of the following forms:

$$q_i S_j S_k q_l, \tag{2.1}$$

$$q_i S_j L q_l, \tag{2.2}$$

$$q_i S_j R q_l. \tag{2.3}$$

Note that $L, R \notin \Sigma$ and $S_j, S_k \in \Sigma \cup \{\sqcup\}$. The quadruple (2.1) specifies that if the machine is in state q_i and the cell that the scanning head scans contains the symbol S_j, then the scanning head replaces S_j by S_k and the machine enters state q_l. The quadruples (2.2) and (2.3) specify

A. Syropoulos, *Theory of Fuzzy Computation*, IFSR International Series on Systems Science and Engineering 31, DOI 10.1007/978-1-4614-8379-3_2,

that if the machine is in state q_i and the cell that the scanning head scans contains the symbol S_j, then the scanning head moves to the cell to the left of the current cell or to the cell to the right of the current cell, respectively, and the machine enters the state q_l. Sometimes the following quadruple is also considered:

$$q_i S_j q_k q_l. \tag{2.4}$$

This quadruple is particularly useful if we want to construct a Turing machine that will compute *relatively computable functions*. These quadruples provide a Turing machine with a means of communicating with an external agency that can give correct answers to questions about a set $A \subset \mathbb{N}$. In particular, when a machine is in state q_i and the cell that the scanning head scans contains the symbol S_j, then the machine can be thought of asking the question, "Is $n \in A$?" Here n is the number of S_1's that are printed on the tape. If the answer is "yes," then the machine enters state q_k; otherwise it enters state q_l. Turing machines equipped with such an external agency are called *oracle machines*, and the external agency is called an *oracle*.

Turing machines are used to compute the value of functions $f(n_1, ..., n_m)$ that take values in $\mathbb{N}^m$. Each argument $n_i \in \mathbb{N}$ is represented on the tape by preprinting the symbol S_1 on $n_i + 1$ consecutive cells. Typically, such a block of cells is denoted by $\overline{n_i}$. Argument representations are separated by a blank cell (i.e., a cell on which the symbol $\sqcup$ is printed), while all other cells are empty (i.e., the symbol S_0 has been preprinted on each cell). It is customary to represent such a block of cell with the expression

$$\overline{(n_1, n_2, ..., n_m)} = \overline{n_1} \sqcup \overline{n_2} \sqcup ... \sqcup \overline{n_m}.$$

If α is an expression, then $\langle \alpha \rangle$ will denote the number of S_1 contained in α. In addition,

$$\overline{\langle m - 1 \rangle} = m \quad \text{and} \quad \langle \alpha\beta \rangle = \langle \alpha \rangle + \langle \beta \rangle.$$

It is also customary to use the symbol 1 for S_1. Thus, the sequence $3, 4, 2$ will be represented by the following three blocks of 1's:

$$1111 \sqcup 11111 \sqcup 111.$$

The machine starts at state q_0 and the scanning head is placed atop the leftmost 1 of a sequence of n blocks of 1's. If the machine has reached a situation in which none or more than one quadruple is applicable, the machine halts. Once the machine has terminated, the result of the computation is equal to the number of cells on which the symbol S_1 is printed.

Although the description presented so far is quite formal for my own taste, still fuzzy versions of the Turing machine are extensions of the "standard" formal definition that is given below.

Definition 2.1.2 A Turing machine $\mathcal{M}$ is an octuple $(Q, \Sigma, \Gamma, \delta, \sqcup, \triangleright, q_0, H)$, where

- Q is a finite set of states.

- Σ is the input alphabet.

- Γ is an alphabet, called *working alphabet*, where $\Sigma \subseteq \Gamma$.

The Turing machine's scanning head moves back and forth along the tape. The number that the scanning head displays is its current state, which changes as it proceeds.

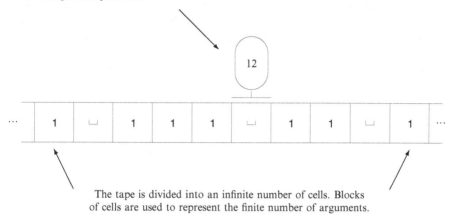

The tape is divided into an infinite number of cells. Blocks of cells are used to represent the finite number of arguments.

Figure 2.1: A typical Turing machine

- $\sqcup \in \Gamma$ is the blank symbol.

- $\triangleright \in \Gamma$ is the *left end symbol*.

- $q_0 \in Q$ is the initial state.

- $H \subseteq Q$ is the set of halting or accepting and rejecting states.

- δ, the *transition* function, is a function from $(Q \setminus H) \times \Gamma$ to $Q \times \Gamma \times \{L, R, N\}$ such that $\mathcal{M}$ may perform an instruction $(q', Y, D) \in Q \times \Gamma \times \{L, R, N\}$, if $\mathcal{M}$ is in state q, the scanning head has just read the symbol X, and $\delta(q, X) = (q', Y, D)$. Depending on the value of D, the machine will move to the left (L), to the right (R), or it will stay still when $D = N$ and the scanning head will overwrite X with Y. Also, $\delta(q, \triangleright) = (q, \triangleright, R)$, which means that whenever the scanning head has read the symbol $\triangleright$, it immediately moves to the right.

Note that here I defined a machine that has a unidirected tape and not a bidirectional tape. One can get the bidirectional version by eliminating all references to the $\triangleright$ symbol.

A *configuration* of $\mathcal{M}$ is an element from

$$C(\mathcal{M}) = \{\triangleright\}\Gamma^*Q\Gamma^+ \cup Q\{\triangleright\}\Gamma^*,$$

where AB denotes strings that are formed by concatenating a string that belongs to A with a string that belongs to B; also, A^* is the set of all finite words over A,[1] that is,

$$A^* = \{\varepsilon\} \cup A \cup (A \times A) \cup (A \times A \times A) \cup \cdots,$$

---

1. Alternatively known as the *free monoid with base A*.

ε is the empty word, and $A^+ = A^* \setminus \{\varepsilon\}$. If $w_1 q a w_2$ is a configuration, then $w_1 \in \{\triangleright\}\Gamma^*$, $w_2 \in \Gamma^*$, $a \in \Gamma$, and $q \in Q$. Moreover, this configuration means that a machine $\mathcal{M}$ is in state q, the content of the tape is $\triangleright w_1 a w_2 \sqcup \sqcup \sqcup ...$, and the scanning head sits atop the $(|w_1| + 1)$-th cell, where $|s|$ is the length of string s. The *initial* configuration is $q_0 \triangleright x$, where x is the input fed to the machine. A configuration whose state component is in H is called a *halted* configuration.

A *step* is a relation $\vdash_{\mathcal{M}}$ on the set of configurations defined as follows:

(i) $...x_{i-1} q x_i x_{i+1} ... \vdash_{\mathcal{M}} ...x_{i-1} q' y x_{i+1} ...$, if $\delta(q, x_i) = (q', y, N)$;

(ii) $...x_{i-1} q x_i x_{i+1} ... \vdash_{\mathcal{M}} ...x_{i-2} q' x_{i-1} y x_{i+1} ...$, if $\delta(q, x_i) = (q', y, L)$;

(iii) $...x_{i-1} q x_i x_{i+1} ... \vdash_{\mathcal{M}} ...x_{i-1} y q' x_{i+1} ...$, if $\delta(q, x_i) = (q', y, R)$; and

(iv) $...x_{n-1} q x_n \vdash_{\mathcal{M}} ...x_{n-1} y q' \sqcup$, if $\delta(q, x_n) = (q', y, R)$.

A *computation* by $\mathcal{M}$ is a sequence of configurations $C_0, C_1, ..., C_n$, for some $n \geq 0$ such that

$$C_0 \vdash_{\mathcal{M}} C_1 \vdash_{\mathcal{M}} C_2 \vdash_{\mathcal{M}} \cdots \vdash_{\mathcal{M}} C_n.$$

A computation from C_0 to C_n can be written compactly as $C_0 \vdash^*_{\mathcal{M}} C_n$.

A computation by $\mathcal{M}$ on input x is a series of actions that start at configuration $C_0 = q_0 \triangleright x$ and is either *infinite* (i.e., nonterminating) or stops at configuration $w_1 q w_2$, where $q \in H$. Assume that $H = \{q_a, q_r\}$, where q_a is the *accepting* state and q_r is the *rejecting* state. Then a computation is called *accepting* if it finishes in the configuration $w_1 q_a w_2$ and *rejecting* if it finishes in the configuration $w_1 q_r w_2$. Furthermore, if a computation on input x is accepting or rejecting, we say that the corresponding machine *accepts* or *rejects* x, respectively. More generally, the words accepted by $\mathcal{M}$ form a formal language $L(\mathcal{M})$.

Typically, a *formal language*, or simply a language, over an *alphabet* Σ is subset of Σ^*, which is often defined by means of a formal *grammar*.

Definition 2.1.3 A grammar is defined to be a quadruple $G = (V_N, V_T, S, \Phi)$ where V_T and V_N are disjoint sets of terminal and nonterminal (syntactic class) symbols, respectively; S, a distinguished element of V_N, is called the starting symbol. Φ is a finite nonempty relation from $(V_T \cup V_N)^* V_N (V_T \cup V_N)^*$ to $(V_T \cup V_N)^*$. In general, an element (α, β) is written as $\alpha \to \beta$ and is called a production or rewriting rule [134].

The formal language *decided* by some Turing machine $\mathcal{M}$ is defined by

$$L(\mathcal{M}) = \left\{ w \mid (w \in \Sigma^*) \wedge \left(\mathcal{M} \text{ accepts } w \right) \right\}.$$

$\mathcal{M}$ *decides* a language L if for any word $w \in \Sigma^*$ either $w \in L$ and $\mathcal{M}$ *accepts* w or $w \notin L$ and $\mathcal{M}$ *rejects* w. A language is called *recursive* or *decidable* if there is a Turing machine that decides it. Also, $\mathcal{M}$ *semidecides* a language L when the machine halts for input w if and only if $w \in L$. A language L is *recursively enumerable* or *semidecidable* if there is a Turing machine that semidecides it.

$\mathcal{M}$ computes a function $F : \Sigma^* \to \Gamma^*$ if

$$\text{for all } x \in \Sigma^* : q_0 \triangleright x \vdash^*_{\mathcal{M}} q_a \triangleright F(x).$$

If $F : \Sigma^* \to \Gamma^*$ is computable by a Turing machine, it is called *recursive*.

Let $\mathcal{M}$ be a Turing machine and let

$$\Psi^{(n)}_{\mathcal{M}}(x_1, x_2, ..., x_n)$$

be a partial function of n arguments. Then, $\mathcal{M}$ computes $\Psi^{(n)}_{\mathcal{M}}$ if for each n-tuple of arguments, $\mathcal{M}$ halts after a finite number of steps. If $\mathcal{M}$ does not terminate on a tuple $(k_1, ..., k_n)$, then $\Psi^{(n)}_{\mathcal{M}}$ is undefined on this tuple. $\mathcal{M}$ computes f if for all $(x_1, ..., x_n)$, $\Psi^{(n)}_{\mathcal{M}}(x_1, ..., x_n)$ is defined and equal to $f(x_1, ..., x_n)$. Now, it is possible to construct a Turing machine $\mathcal{M}'$ that will have as input the description of the controlling device of another Turing machine $\mathcal{M}$ and its arguments. Clearly, both the description of the controlling device and the arguments of the machine have to be encoded. In order to encode the various symbols, one may start by using an injective mapping (i.e., one that preserves distinctness) from the class of symbols into the integers. Thus, each symbol is identified with an integer "label." More generally, sentences are mapped to a unique integer by employing similar methods. Mappings like this are called *codings* and the labels are called *code numbers*. The most common coding is the Gödel numbering, which was invented by Gödel, and the details of this coding are given below.

Suppose that we associate with each basic symbol of a Turing machine an odd number greater than or equal to 3 as follows:

3	5	7	9	11	13	15	17	19	21	...
↑	↑	↑	↑	↑	↑	↑	↑	↑	↑	
R	L	S_0	q_0	S_1	q_1	S_2	q_2	S_3	q_3	...

For each i, S_i is associated with $4i + 7$ and q_i is associated with $4i + 9$. In order to define the encoding of a Turing machine, first we need to define the encoding of an expression and then the encoding of a sequence of expressions.

Definition 2.1.4 Assume that M is a string of symbols $\gamma_1, \gamma_2, ..., \gamma_n$ and that $a_1, a_2, ..., a_n$ are the corresponding integers associated with these symbols. The Gödel number of M is the integer

$$\text{Gn}(M) = \prod_{k=1}^{n} \left(\text{Pr}(k) \right)^{a_k},$$

where $\text{Pr}(k)$ is the kth prime number in order of magnitude.

Example 2.1.1 If $M = q_1 S_0 S_2 q_1$, then $\text{Gn}(M) = 2^{13} \cdot 3^7 \cdot 5^{15} \cdot 7^{13}$, that is,

$$\text{Gn}(M) = 52{,}974{,}066{,}440{,}027{,}250{,}000{,}000{,}000{,}000.$$

Definition 2.1.5 Suppose that $M_1, M_2, ..., M_n$ is a finite sequence of expressions. Then the Gödel number of this sequence is the integer:

$$\prod_{k=1}^{n} \left(\text{Pr}(k) \right)^{\text{Gn}(M_k)}.$$

Definition 2.1.6 Assume that $M_1, M_2,...,M_n$ is any arrangement of the quadruples of a Turing machine $\mathcal{M}$ without repetitions. Then the Gödel number of the sequence $M_1, M_2,...,M_n$ is a Gödel number of $\mathcal{M}$.

Clearly, a Turing machine consisting of n quadruples has $n!$ different Gödel numbers.

Definition 2.1.7 A universal Turing machine $\mathcal{U}$ is a Turing machine that can be employed to compute any function of one argument that an ordinary Turing machine $\mathcal{M}$ can compute.

Practically, this means that given a Turing machine $\mathcal{M}$ with a Gödel number m that computes the function $f(x)$, then

$$\Psi_{\mathcal{U}}^{(2)}(m, x) = f(x) = \Psi_{\mathcal{M}}^{(1)}(x).$$

Thus, if the number m is written on the tape of $\mathcal{U}$, followed by the number x, $\mathcal{U}$ will compute the number $\Psi_{\mathcal{M}}^{(1)}(x)$. Also, the universal Turing machine can be used to compute functions with n arguments, but I am not going to describe how this can be done (see [39] for the relevant details).

Function $\Psi_{\mathcal{U}}^{(2)}$ is just an example of a function that has as arguments a "program" and its "input." Another interesting example of such a function is the so-called halting function:

$$h(m, x) = \begin{cases} 1, & \text{when } \mathcal{M} \text{ starts with input } x \text{ and eventually stops,} \\ 0, & \text{otherwise,} \end{cases}$$

where m is the Gödel number of $\mathcal{M}$. Whether this function is computable is equivalent to the halting problem. This, in turn, can be summarized as follows: is there an *effective* procedure such that given any m and any x we can determine whether $\Psi_{\mathcal{U}}^{(2)}(m, x)$ is defined or not?

It can be shown using the *diagonalization argument* that there is no such method. In particular, by using this principle one can easily prove that no Turing machine can determine whether $\Psi_{\mathcal{U}}^{(2)}(m, x)$ is defined. But what exactly is this argument?

The diagonalization argument was devised by Georg Ferdinand Ludwig Philip Cantor to show that the set of real numbers is not countably infinite. It is based on the diagonalization principle, a fundamental proof technique, which can be stated as follows:

Principle 2.1.1 (Diagonalization Principle) Assume that R is a binary relation on a set A. Also, assume that D, the diagonal set for R, is the set

$$\Big\{ a \mid (a \in A) \wedge \big((a, a) \notin R\big) \Big\}.$$

For each $a \in A$, suppose that $R_a = \{b \mid (a, b) \in R\}$. Then D is distinct from each R_a.

Paulo Cotogno [38] has presented a simplified proof of the insolvability of the halting problem: Assume that $\Psi_1^{(1)}, \Psi_2^{(1)}, \Psi_3^{(1)},...$ is an enumeration of the computable (partial) functions. In addition, let us define the halting function

$$f(x, y) = \begin{cases} 1, & \text{if } \Psi_x^{(1)}(y) \text{ converges,} \\ 0, & \text{if } \Psi_x^{(1)}(y) \text{ does not converge,} \end{cases}$$

and the diagonal monadic function

$$g(x) = \begin{cases} 1, & \text{when } f(x,x) = 0, \\ \perp, & \text{when } f(x,x) = 1. \end{cases}$$

Here $\perp$ denotes an *undefined* or *divergent* value, depending on the context. Suppose that $g(x)$ is computable. Then there is an i such that $g(x) = \Psi_i^{(1)}(x)$. This implies that $\Psi_i^{(1)}(i) = 0$, that is, $g(i) = 0$. The last equation is equivalent to $f(i,i) = 0$ and this implies that $\Psi_i^{(1)}(i)$ is undefined, which is obviously a contradiction.

The interesting thing about this proof is that it holds for all definitions of computability. In other words, if one considers a class of machines, then it is impossible for them to compute their own halting functions. Indeed, most models of computation with *hypercomputational*[2] capabilities have been examined by Toby Ord and Tien Kieu [97], who have concluded that none of them is able to solve its own halting problem. However, this does not mean that some machine cannot solve the halting problem for Turing machines or some hypermachines— there is no logical inconsistency here.

2.2   Extensions of Turing Machines

In this section, I will briefly present three standard extensions of the Turing machine: nondeterministic, probabilistic, and multitape Turing machines. It has been argued that any probabilistic Turing machine can be transformed into a nondeterministic one by ignoring the probabilities. However, in the lights of the discussion in Sect. 1.2, where it was argued that nondeterminism is primitive notion, while randomness is not, this "equivalence" is meaningless.

2.2.1   Nondeterministic Turing Machines

Quite naturally, Turing [135, p. 232] was the first to describe an extension to his standard model of computation. He called these extended Turing machines *c-machines* and described them as follows:

> For some purposes we might use machines (choice machines or c-machines) whose motion is only partially determined by the configuration (hence the use of word "possible" in §1). When such a machine reaches one of these ambiguous configurations, it cannot go on until some arbitrary choice has been made by an external operator. This would be the case if we were using machines to deal with axiomatic systems.

In a sense, one could argue that c-machines correspond to machines that function in a nondeterministic way; nevertheless, there is a precise formulation of the latter:

Definition 2.2.1 An octuple $(Q, \Sigma, \Gamma, \Delta, \sqcup, \triangleright, q_0, H)$ is a *nondeterministic Turing machine* $\mathcal{N}$ if $\Delta : (Q \setminus H) \times \Gamma \to \mathscr{P}(Q \times \Gamma \times \{L, R, N\})$ is the transition function and all others are as for the standard Turing machine.

---

2. In a nutshell, hypercomputation is the idea that there are conceptual and real computing machines that transcend the capabilities of the Turing machine. See [126] for an overview of the field of hypercomputation.

A configuration of $\mathcal{N}$ is an element from

$$C(\mathcal{N}) = \{\triangleright\}\Gamma^*Q\Gamma^* \cup Q\{\triangleright\}\Gamma^*.$$

The configuration $q_0 \triangleright w$ is the *initial configuration of $\mathcal{N}$ on w*, where w is the input fed to the machine. A configuration whose state component is in H is called a halted configuration.

A step of $\mathcal{N}$ is a binary relation $\vdash_{\mathcal{N}} \subseteq C(\mathcal{N}) \times C(\mathcal{N})$ that is defined as follows:

(i) $x_1 x_2 ... x_{i-1} q x_i x_{i+1} ... x_n \vdash_{\mathcal{N}} x_1 x_2 ... x_{i-1} q' y x_{i+1} ... x_n$, if $(q', y, N) \in \Delta(q, x_i)$;

(ii) $x_1 x_2 ... x_{i-1} q x_i x_{i+1} ... x_n \vdash_{\mathcal{N}} x_1 x_2 ... x_{i-2} q' x_{i-1} y x_{i+1} ... x_n$, if $(q', y, L) \in \Delta(q, x_i)$;

(iii) $x_1 x_2 ... x_{i-1} q x_i x_{i+1} ... x_n \vdash_{\mathcal{M}} x_1 x_2 ... x_{i-1} y q' x_{i+1} ... x_n$, if $(q', y, R) \in \Delta(q, x_i)$; and

(iv) $x_1 x_2 ... x_{n-1} q x_n \vdash_{\mathcal{M}} x_1 x_2 ... x_{n-1} y q' \sqcup$, if $(q', y, R) \in \Delta(q, x_i)$.

A computation of $\mathcal{N}$ is a sequence $C_0, C_1, ..., C_n$ of configurations such that

$$C_i \vdash_{\mathcal{N}} C_{i+1},$$

for $i = 0, 1, 2, ..., n$. A *computation of $\mathcal{N}$ on an input x* is any computation starting from the initial configuration and either halts, in which case it halts in a configuration $w_1 q w_2$, or is infinite. When a machine halts in an accepting/rejecting state, we say that $\mathcal{N}$ accepts/rejects the input word x. The language $L(\mathcal{N})$ accepted by the nondeterministic Turing machine $\mathcal{N}$ is

$$L(\mathcal{N}) = \left\{ x \mid \left(x \in \Sigma^* \right) \wedge \left(q_0 \triangleright x \vdash_{\mathcal{N}} y q_a z \quad \text{for some } y, z \in \Gamma^* \right) \right\}$$

$$= \left\{ x \mid \left(x \in \Sigma^* \right) \wedge \left(\mathcal{N} \text{ accepts } x \right) \right\}.$$

2.2.2 Probabilistic Turing Machines

Santos [112] introduced the *probabilistic Turing machine* as a probabilistic extension of ordinary Turing machines.

Definition 2.2.2 A probabilistic Turing machine P is a triple (S, Q, p), where S is a set of symbols that the machine can print, Q is the set of internal states, $S \cap Q = \emptyset$, and $p : Q \times S \times V \times Q \to [0, 1]$, where $V = S \cup \{R, L, N\}$ and $R \notin S$, $L \notin S$, and $N \notin S$. Function p satisfies the following conditions:

(i) $\sum_{v \in V} \sum_{q' \in Q} p(q, s, v, q') = 1$ for every $q \in Q, s \in S$; and

(ii) for every $s \in S$, $p(q, s, N, q') = 0$ if $q \neq q'$.

Given that a machine is at state q and the symbol just scanned is s, the value of function p is the conditional probability of the "next act."

Definition 2.2.3 Assume that $Z = (S, Q, p)$ is a probabilistic Turing machine. Then $\alpha \in (S \cup Q)^*$ is an *expression* of Z. α is an instantaneous expression of Z if and only if it contains exactly one $q \in Q$ and q is not the rightmost symbol. α is a tape expression if and only if it consists entirely of symbols from S. When α is an instantaneous expression of Z that contains $q \in Q$ and u is the symbol immediately to the right of q, then q is called the state of Z at α and u the symbol scanned by Z at α. The tape expression obtained by removing q from α is called the expression on the tape of Z at α.

Definition 2.2.4 Assume that $Z = (S, Q, p)$ is a probabilistic Turing machine. Then, for every instantaneous expression α and β of Z, define

$$
\mathsf{q}_Z(\alpha, \beta) = \begin{cases}
p(q, s, s', q') & \text{if } \alpha = \gamma q s \delta, \quad \beta = \gamma q' s' \delta, \quad s' \in S, \\
p(q, s, R, q') & \text{if } \alpha = \gamma q s s' \delta, \quad \beta = \gamma s q' s' \delta, \quad s' \in S, \\
& \text{or } \alpha = \gamma q s, \quad \beta = \gamma s q' \llcorner, \\
p(q, s, L, q;) & \text{if } \alpha = \gamma s' q s \delta, \quad \beta = \gamma q' s' s \delta, \quad s' \in S, \\
& \text{or } \alpha = q s \delta, \quad \beta = q' \llcorner s \delta, \\
0 & \text{otherwise,}
\end{cases}
$$

where $\mathsf{q}_Z(\alpha, \beta)$ is the probability that when the current instantaneous expression of Z is α, the next one will be β; γ and δ are (possibly empty) tape expressions of Z.

One can extend $\mathsf{q}_Z(\alpha, \beta)$ as follows:

$$
\begin{aligned}
\mathsf{q}_Z^0(\alpha, \beta) &= 1 \quad \text{if } \alpha = \beta, \\
&= 0 \quad \text{if } \alpha \neq \beta, \\
\mathsf{q}_Z^n(\alpha, \beta) &= \sum_\gamma \mathsf{q}_Z^{n-1}(\alpha, \gamma) \mathsf{q}_Z^0(\gamma, \beta),
\end{aligned}
$$

where the summation ranges over all instantaneous expressions γ. Also, $\mathsf{q}_Z^n(\alpha, \beta)$ can be viewed as the probability that instantaneous expression of Z will be β *after n steps* provided that Z *starts* with instantaneous expression α. It can be proved that for every instantaneous expression α and nonnegative integer n it holds that

$$
\sum_\beta \mathsf{q}_Z^n(\alpha, \beta) \leq 1.
$$

Definition 2.2.5 Suppose that $Z = (S, Q, p)$ is a probabilistic Turing machine. Then, for all instantaneous expressions α and β of Z and for all $n = 1, 2, \dots$ define

$$
t_Z^{(n)}(\alpha, \beta) = p(q, s, N, q) \mathsf{q}_Z^{(n-1)}(\alpha, \beta),
$$

where q is the state of Z at β and s is the symbol scanned by Z when at β. Furthermore, define

$$
t_Z(\alpha, \beta) = \sum_{n=1}^\infty t_Z^{(n)}(\alpha, \beta),
$$

where $t_Z^{(n)}(\alpha, \beta)$ can be viewed as the probability that the machine starts with instantaneous expression α and terminates with instantaneous expression β in n steps. The interpretation of $t_Z^{(n)}(\alpha, \beta)$ is obvious.

Definition 2.2.6 A k-ary *random* function ϕ is a function from the collection of all $(k+1)$-tuples of nonnegative integers to $[0,1]$ that satisfies the following inequality:

$$\sum_{m=0}^{\infty} \phi(m_1, m_2, ..., m_k, m) \le 1$$

for all k-tuples $(m_1, m_2, ..., m_k)$.

Definition 2.2.7 Suppose that $Z = (S, Q, p)$ is a probabilistic Turing machine. Then, all positive integers k are associated with a k-ary random function $\Phi_Z^{(k)}$ as follows:

$$\Phi_Z^{(k)}(m_1, m_2, ..., m_k, m) = \sum_{\langle \beta \rangle = m} t_Z(\alpha, \beta),$$

where

$$\alpha = q_1 \overline{(m_1, m_2, ..., m_k)}, \quad q_1 \in Q$$

and the summation ranges over all instantaneous expressions β of Z such that $\langle \beta \rangle = m$.

Here q_1 plays the role of the initial state of a stochastic sequential-like machine. However, if this initial state is not given, but, instead, the initial distribution is given, then

$$\Phi_Z^{(k)}(m_1, m_2, ..., m_k, m) = \sum_{\langle \beta \rangle = m} \sum_{i=1}^{|Q|} h(q_i) t_Z(\alpha_i, \beta),$$

where

$$\alpha_i = q_i \overline{(m_1, m_2, ..., m_k)}, \quad \text{and} \quad q_i \in Q$$

and $|Q|$ is the cardinality of Q.

Definition 2.2.8 A k-ary random function ϕ is computable if and only if $\phi = \Phi_Z^{(k)}$ for some probabilistic Turing machine Z.

Santos [114] generalized his definition of a probabilistic Turing machine as follows:

Definition 2.2.9 A probabilistic Turing machine is specified by a quintuple (A, B, Q, p, h) where

- A is the printing alphabet;

- B is the auxiliary alphabet;

- Q is the set of internal states;

- $p : Q \times U \times V \times Q \to [0,1]$, where $U = A \cup B$, $V = U \cup Q \cup \{R, L, N\}$, $R \notin U \cup Q$, $L \notin U \cup Q$, and $N \notin U \cup Q$, gives the probability of the next action; and

- $h : Q \to [0,1]$ is the probability that the initial state is q.

Table 2.1: The definition of $q_{Z,T}(\alpha, \beta)$ (see Definition 2.2.11), where $\gamma, \delta \in (A \cup B)^*$, $q, q' \in Q$, and $u, u' \in A \cup B$

$$
\begin{aligned}
q_{Z,T}(\alpha, \beta) \quad &= p(q, u, u', q') && \text{if } \alpha = \gamma q u \delta, \beta = \gamma q' u' \delta, u' \neq u; \\
&= p(q, u, R, q') && \text{if } \alpha = \gamma q u u' \delta, \beta = \gamma u q' u' \delta, \gamma u \neq \text{\textvisiblespace}, \\
& && \text{or } \alpha = q u u' \delta, \beta = q' u' \delta, u = \text{\textvisiblespace}, \\
& && \text{or } \alpha = \gamma q u, \beta = \gamma u q' \text{\textvisiblespace}, \gamma u \neq \text{\textvisiblespace}, \\
& && \text{or } \alpha = q u, \beta = q' \text{\textvisiblespace}, u = \text{\textvisiblespace}; \\
&= p(q, u, L, q') && \text{if } \alpha = \gamma u' q u \delta, \beta = \gamma q' u' u \delta, u \delta \neq \text{\textvisiblespace}, \\
& && \text{or } \alpha = \gamma u' q u, \beta = \gamma q' u', u = \text{\textvisiblespace}, \\
& && \text{or } \alpha = q u \delta, \beta = q' \text{\textvisiblespace} u \delta, u \delta \neq \text{\textvisiblespace}, \\
& && \text{or } \alpha = q u, \beta = q' \text{\textvisiblespace}, u = \text{\textvisiblespace}; \\
&= p(q, u, u, q') + && \text{if } \alpha = \gamma q u \delta, \beta = \gamma q' u \delta; \\
& \quad \sum_{q'' \in Q} p(q, u, q', q'') T(\langle \alpha \rangle) + \\
& \quad \sum_{q'' \in Q} p(q, u, q'', q')[1 - T(\langle \alpha \rangle)] \\
&= 0 && \text{otherwise}
\end{aligned}
$$

Functions p and h must satisfy the following conditions:

(i) for all $q \in Q, u \in U, \sum_{v \in V} \sum_{q' \in Q} p(q, u, v, q') = 1$; and

(ii) $\sum_{q \in Q} h(q) = 1$.

Function p is called the *transition* function and h is called the *initial distribution*. When for a particular $q_0 \in Q$ it holds that $h(q_0) = 1$ while $h(q) = 0$ for $q \neq q_0$, then q_0 is the initial state.

A probabilistic machine is *deterministic* if and only if the range of p and h is the set $\{0, 1\}$. Also, a machine is called *simple* if and only if $p(q, u, v, q') = 0$ for all $q, q' \in Q, u \in A \cup B$, and $v \in Q$.

Remark 2.2.1 An ordinary Turing machine is a deterministic probabilistic Turing machine while machines of the type of Definition 2.2.2 are simple machines.

Given a machine Z an expression of this machine is an element of $(A \cup B \cup Q)^*$. Also, a word is an element of A^*.

Definition 2.2.10 Assume that Z is a probabilistic Turing machine. Then an expression α of Z is an instantaneous description of Z if and only if

(i) α contains exactly one $q \in Q$ and q is not the rightmost symbol of α;

(ii) the leftmost symbol of α is not $\text{\textvisiblespace}$; and

(iii) the rightmost symbol of α is not $\text{\textvisiblespace}$ unless it is the symbol immediately to the right of q.

The collection of all instantaneous descriptions of some machine Z will be denoted by $\mathscr{I}(Z)$. Also, if α is an expression of some machine Z, then $\langle \alpha \rangle$ will denote the word of Z obtained by removing from α all symbols that do not belong to A provided that α contains symbols from A; otherwise $\langle \alpha \rangle = \sqcup$.

Definition 2.2.11 Assume that $Z = (A, B, Q, p, h)$ is a probabilistic Turing machine and T a random set[3] in A^*. For every $\alpha, \beta \in \mathscr{I}(Z)$, $q_{Z,T}(\alpha, \beta)$ is defined as in Table 2.1 in p. 23.

The expression $q_{Z,T}(\alpha, \beta)$ denotes the probability that the *next* instantaneous description of Z relative to T will be β provided that Z *starts* with the instantaneous description α.

Definition 2.2.12 For all $\alpha, \beta \in \mathscr{I}(Z)$ and $n = 0, 1, 2, \ldots$ define $q_{Z,T}^{(i)}$ as follows:

$$q_{Z,T}^{(0)}(\alpha, \beta) = 1 \text{ if } \alpha = \beta,$$
$$q_{Z,T}^{(0)}(\alpha, \beta) = 0 \text{ if } \alpha \neq \beta,$$
$$q_{Z,T}^{(n)}(\alpha, \beta) = \sum_{\gamma \in \mathscr{I}(Z)} q_{Z,T}^{(n-1)}(\alpha, \gamma) q_{Z,T}(\gamma, \beta).$$

The expression $q_{Z,T}^{(n)}$ is the probability that the instantaneous description of Z relative to T will be β *after n steps* provided that Z *starts* with α.

Definition 2.2.13 For all $\alpha, \beta \in \mathscr{I}(Z)$ and $n = 1, 2, \ldots$, define

$$t_{Z,T}^{(n)}(\alpha, \beta) = p(q, u, N, q) q_{Z,T}^{(n-1)}(\alpha, \beta),$$

where q is the state of Z at β and u the symbol scanned by Z at β. In addition, define

$$t_{Z,T}(\alpha, \beta) = \sum_{n=1}^{\infty} t_{Z,T}^{(n)}(\alpha, \beta).$$

The expression $t_{Z,T}^{(n)}(\alpha, \beta)$ denotes the probability that Z will *terminate* with β relative to T *after n steps* provided that Z *starts* with α. Also, $t_{Z,T}(\alpha, \beta)$ is the probability that Z will *terminate* with β relative to T *after a finite number of steps* given that Z *starts* with α.

The following definitions are needed for the exposition in Sect. 4.7. Assume that $\mathscr{A}$ is the set $\{a_1, a_2, \ldots, a_m\}$, where $m > 0$ and $a_1 = 1$, and that T is a random set in $\mathscr{A}^*$.

Definition 2.2.14 Function $\mathscr{N} : \mathscr{A}^* \to \mathbb{N}$ is defined as follows:

$$\begin{aligned} \mathscr{N}(w) &= \sum_{i=1}^{n} k_i m^{i-1} \quad \text{if } w = a_{k_n} a_{k_{r-1}} \ldots a_{k_1}, n > 0, \\ &= 0 \quad \text{if } w = \varepsilon. \end{aligned}$$

Conversely, function $\mathscr{W} : \mathbb{N} \to \mathscr{A}^*$ is defined as follows:

$$\begin{aligned} \mathscr{W}(n) &= w \quad \text{if and only if } \mathscr{N}(w) = n, \\ &= \varepsilon \quad \text{when } n < 0. \end{aligned}$$

---

3. A *random* set C in an ordinary set X is characterized by a function $C : X \to [0, 1]$, where $C(x)$ denotes the probability that $x \in X$.

Function $\mathcal{N}$ is called *codifier* and function $\mathcal{W}$ is called *decodifier*. These names are justified by the following lemmata:

Lemma 2.2.1 *There is a simple deterministic probabilistic Turing machine* $Z = (A, B, Q, p, q_0)$ *such that for every T and* $w \in \mathcal{A}^*$

$$t_{Z,T}(q_0 w, \beta) \quad = 1 \quad if \beta = q \mathcal{N}(w),$$
$$= 0 \quad otherwise.$$

Machine Z will be called the coding machine with final state q.

Lemma 2.2.2 *There is a simple deterministic probabilistic Turing machine* $Z = (A, B, Q, p, q_0)$ *such that for every T and nonnegative n*

$$t_{Z,T}(q_0 \overline{n}, \beta) \quad = 1 \quad if \beta = q \mathcal{W}(n),$$
$$= 0 \quad otherwise,$$

where q is a terminating state of Z.

Machine Z will be called the decoding machine with final state q.

2.2.3   Multitape Turing Machines

A multitape Turing machine is one that has several tapes, where each tape has its own scanning head and all of them are connected to the unique controlling device. At any given moment, the machine is in a unique state. The scanning heads conclude reading the symbols scanned in a single step. Depending on the current state and these symbols, the machine rewrites some cells or moves some scanning heads to the left or the right while it changes its state. Obviously, a multitape machine with just one tape is an ordinary Turing machine.

Definition 2.2.15 Assume that k is a positive integer. Then, a *k-tape Turing machine* is an octuple $(Q, \Sigma, \Gamma, \delta, \llcorner, \triangleright, q_0, H)$, where with exception of δ, the transition function, all other components are as in the definition of the ordinary Turing machine (see Definition 2.1.2). The transition function is a function from $(Q \setminus H) \times \Gamma^k$ to $Q \times (\Gamma \times \{L, R, N\})^k$. That is, for each state q and each k-tuple of tape symbols $(a_1, ..., a_k)$, $\delta(q, (a_1, ..., a_k)) = (q', (b_1, ..., b_k))$, where q' is the new state and each b_j is the "action" taken by the machine on tape j.

Computation takes place in all tapes of a multitape Turing machine. Therefore, a configuration of such a machine must include information about all tapes:

Definition 2.2.16 A *configuration* of a multitape Turing machine $\mathcal{M} = (Q, \Sigma, \Gamma, \delta, \llcorner, \triangleright, q_0, H)$ with k tapes is an element from

$$C(\mathcal{M}) = (\{\triangleright\}\Gamma^*)^k Q(\Gamma^+)^k \cup Q(\{\triangleright\}\Gamma^*)^k.$$

2.3   The "Most General" Approach to the Concept of an Algorithm

In 1953, Andrei Nikolaevich Kolmogorov [76] presented the sketch of a new definition of the notion of algorithm. Later on, a complete formulation was presented in a paper co-authored by Kolmogorov and Vladimir Andreevich Uspensky [78]. The basic idea behind this formulation is that an algorithm Γ is applied to a "condition" $A \in \mathfrak{A}(\Gamma)$, where $\mathfrak{A}(\Gamma) \subset \mathfrak{S}(\Gamma)$, $\mathfrak{A}(\Gamma)$ is the class of initial states, and $\mathfrak{S}(\Gamma)$ is the class of possible states, to get a "solution" B that is extracted from the terminal state $T \in \mathfrak{C}(\Gamma)$, where $\mathfrak{C}(\Gamma)$ is the class of terminal states. Function Ω_Γ takes a state and generates the next state. Thus, a computation is a series of state transformations: $A_1 = \Omega_\Gamma(A)$, $A_2 = \Omega_\Gamma(A_1)$,..., $A_n = \Omega_\Gamma(A_{n-1})$. The process stops when either a terminal state has been reached or when a state cannot be handled by function Ω_Γ. However, it is quite possible that the process may never terminate. Now what is really interesting is that Grigoriev [61] found a way to show that Kolmogorov's notion of algorithm is stronger than Turing machines. In different words, Grigoriev found an algorithm à la Kolmogorov–Uspensky that has hypercomputational properties. In what follows, I will briefly present algorithms à la Kolmogorov–Uspensky, in general, and Grigoriev's algorithm, in particular.

2.3.1   States

A state S is defined in terms of a finite number of elements $\bigcirc_1,...,\bigcirc_k$, where each $\bigcirc_i$ has one of the types $T_0, T_1,...,T_n$, and in terms of groups of relations having one of the types $R_1,...,R_m$. For each type of relation R_i only a fixed number k_i of related elements will be considered. Note that the number of elements, types, and relations are fixed for any algorithm and cannot be changed. In general, each relation R_j is not symmetric. Thus, the order in which elements are related plays an important role for states. Also, the index of the type of each element $\bigcirc_j$ is placed inside the circle as in $\textcircled{1}_j$. As for the indices adjoining the circles, they have nothing to do with the structure of the state.

The notion of a state depends on the presentation of a definite order of all relations in which each element occurs. Thus, the links between elements $\bigcirc_i$ and a relation of type R can be visualized by the following schema:

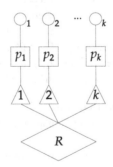

Here p_j indicates the "area" that a particular relation of type R occupies in the ordered sequence of relations in which the element $\bigcirc_j$ is involved. By introducing additional types of elements

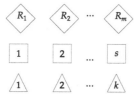

one can replace old algorithms with new ones that differ from the old ones in the way we express states. In particular, there will be only one type of relation between two elements. For example, such a relation between elements $\bigcirc_1$ and $\bigcirc_2$ will be represented as follows:

A direct consequence of this transformation is that all elements related to a particular element will be of *different* types. In fact, this consequence is a condition on all elements that are related to some other element.

One has to specify a method by which the "active part" of state is chosen. Here the term "active part" refers to the structure that uniquely determines the transformation that maps S to $S' = \Omega(S)$. More specifically, it is assumed that in the "active part" of a state there is a distinguished *initial* element. Since Kolmogorov wanted to reduce complexity, it was felt that all elements of a state should be linked to the initial element $\bigcirc$ by a chain

$$\bigcirc \!-\!\!-\! \bigcirc^{(1)} \bigcirc^{(2)} \cdots -\!\bigcirc^{(\lambda)}$$

of length $\lambda \leq N$, where N is a constant number, which is associated with each algorithm. In addition, the operation $S' = \Omega(S)$ must rely solely on information about the structure of the "active part" of S. For example, assume that initial elements are of types T_0 and T_1. Then, when a transformation of the initial element from type T_0 to one of type T_1 and this implies that the terminal state has been reached, then this is the simplest case of all possible transformations.

2.3.2   Basic Ideas from Combinatorial Topology

Kolmogorov's concept of an algorithm uses notions from combinatorial topology. So, since these notions are not widely known, I will briefly introduce them. This exposition is based on [106]. In what follows, I assume a basic understanding of Euclidean spaces.[4]

A system of points $x_0, x_1,...,x_n$ of an n-dimensional linear space $\mathbb{R}^n$ is called *independent* if the system of vectors

$$(x_1 - x_0), ..., (x_k - x_0)$$

is linearly independent. Assume that a and b are two distinct points of the Euclidean space $\mathbb{R}^n$. The set of all points $x \in \mathbb{R}^n$ of the form $x = \lambda a + \mu b$, where λ and μ are real numbers such that

$$\lambda + \mu = 1, \quad \lambda \geq 0, \quad \text{and} \quad \mu \geq 0,$$

will be called the *segment* $(a, b) = (b, a)$ with endpoints a and b. A set M of points of the Euclidean space $\mathbb{R}^n$ is called *convex* if $a \in M$ and $b \in M$ imply $(a, b) \in M$.

---

4. Readers not familiar with these notions can consult Springer's "Encyclopedia of Mathematics" web page.

Definition 2.3.1 Suppose that a_0, a_1,...,a_r is a system of independent points of the n-dimensional Euclidean space $\mathbb{R}^n$, $r \leq n$. The set A^r of the points x of the space $\mathbb{R}^n$ of the form

$$x = \lambda^0 a_0 + \lambda^1 a_1 + \cdots + \lambda^r a_r,$$

where λ^0, λ^1,...,λ^r are real numbers that satisfy the conditions

$$\lambda^0 + \lambda^1 + \cdots + \lambda^r = 1 \quad \text{and} \quad \lambda^i \geq 0, \ i = 0, 1, ..., r,$$

is called an *r-dimensional simplex*, or just an *r-simplex*, and is written as $A^r = (a_0, a_1, ..., a_r)$. The points a_0, a_1,...,a_r are called *vertexes* and are contained in the simplex A^r.

Assume that $A^r = (a_0, a_1, ..., a_r)$ is an r-simplex in $\mathbb{R}^n$ and that $a_k = a_{i_k}, k = 0, 1, ..., s, 0 \leq s \leq r$, a subset of the vertexes of A^r. The vertexes a_0, a_1,...,a_r are independent which means that the vertexes a_0, a_1,...,a_s are also independent and so $C^s = (a_0, a_1, ..., a_s)$ is a simplex in $\mathbb{R}^n$. The simplex C^s is called an *s-dimensional face*, or just a *face*, of the simplex A^r. Two simplexes A and B of the Euclidean space $\mathbb{R}^n$ are *properly situated* either if they are nonintersecting or if their intersection $A \cap B$ is a common face of A and B. Two faces of a simplex are always properly situated.

Definition 2.3.2 A finite set K of simplexes of the Euclidean space $\mathbb{R}^n$ is called a *geometric complex*, or just a *complex*, if K satisfies the following conditions:

(i) if A is a simplex of K, then every face of A is also in K; and

(ii) every two simplexes of K are properly situated.

A *subcomplex* of a complex K is any complex L all of those simplexes that are contained in K.

2.3.3 The Concept of an Algorithm

For an algorithm Γ there is an ordered set $\mathfrak{X}$ of classes, unlimited in size, of *elements* T_0, T_1,...,T_n. These elements are used to build the *states*. Also, they are mutually disjoint and their union is denoted by T. As was noted above, the elements of T will be denoted by circles $\bigcirc$. When an element belongs to some class T_j, we may also say that this is an element of type T_j.

In what follows the term complex on a set $\mathfrak{J}$ will refer to an ordinary one-dimensional complex of vertexes from T, that is, the union $K = K_0 \cup K_1$, where $K_0 = \{\bigcirc_i \mid \bigcirc_i \in T\}$ is a finite set of *vertexes* and $K_1 = \{\bigcirc_k\!\!-\!\!\bigcirc_l \mid \bigcirc_k, \bigcirc_l \in K_0\}$ contains certain pairs of elements of K_0 that are called *edges* of the complex K.

The set $\mathfrak{S}_{\mathfrak{J}}$ contains those complexes of K on $\mathfrak{J}$ that have the following properties:

(i) given a fixed vertex, then all vertexes that are joined by edges to this vertex have different types of T_i; and

(ii) there exists only one vertex whose type is either T_0 or T_1 and which is called the *initial vertex*; all other vertexes are of types T_i, where $i \geq 2$.

The set $\mathfrak{A}_3 \subset \mathfrak{S}_3$ contains the initial vertexes of type T_0 and the set $\mathfrak{C}_3 \subset \mathfrak{S}_3$ contains the initial vertexes of type T_1. The complexes of $\mathfrak{S}_3$ are considered the states of the algorithm Γ while the complexes of $\mathfrak{A}_3$ are considered the initial states of Γ and the complexes $\mathfrak{C}_3$ the terminal states of Γ:

$$\mathfrak{S}(\Gamma) = \mathfrak{S}_3, \quad \mathfrak{A}(\Gamma) = \mathfrak{A}_3, \quad \mathfrak{C}(\Gamma) = \mathfrak{C}_3.$$

The active part $U(S)$ of a state S is the subcomplex of the complex S that consists of the vertexes and edges that belong to chains of length $\lambda \leq N$ that contain the initial vertex. Here, N is an arbitrary number, fixed for a given algorithm and a chain is a complex of the form $O_0 \!-\! O_1 \!-\! O_2 \!-\! \cdots \!-\! O_\lambda$. The length of the chain is equal to the number of edges occurring in it. A complex is called connected if any two of its vertexes can be joined by a chain.

The *outer part* $V(S)$ of a state S is the subcomplex that consists of the vertexes not connected with the initial vertex by chains of length $\lambda < N$ and of edges occurring in those chains that contain the initial vertex that have length $\lambda \leq N$. The intersection

$$L(S) = U(S) \cap V(S)$$

is the *boundary* of the active part of the state S. It follows that $L(S)$ is a zero-dimensional complex, that is, it has no edges, and consists of those vertexes that can be joined to the initial vertex by chains of length $\lambda = N$ but, at the same time, they cannot be joined to it by shorter chains.

Remark 2.3.1 The operation $\Omega_\Gamma(S) = S'$ must be such that transforming S into S' should require only the transformation into a new complex of the active part $U(S)$ and no change in the structure of the outer part $V(S)$. In addition, the type of the transformation of $U(S)$ must be determined only by the structure of the complex $U(S)$.

Remark 2.3.2 Two complexes K' and K'' are *isomorphic* if one can put them into one-to-one correspondence in the following way:

(i) vertexes correspond to vertexes and edges to edges;

(ii) an edge that joins the vertexes O'_i and O'_j corresponds to an edge that joins the vertexes O''_i and O''_j; and

(iii) corresponding vertexes have the same type.

It is a fact that the number of non-isomorphic active parts $U(S)$ possible in a given algorithm is bounded. This implies that the transformation rules are easy to formulate in finite form.

An algorithm is transformed according to a set of isomorphic pairs of states:

$$\begin{aligned}
U_1 &\to W_1, \\
U_2 &\to W_2, \\
&\vdots \quad \vdots \\
U_k &\to W_k.
\end{aligned}$$

For each pair i there is a mapping φ_i that "sends" the subcomplex $L(U_i)$ to the subcomplex $L'(W_i)$. If for the vertexes $\bigcirc'_\mu \in L(U_i)$ and $\bigcirc_v \in L'(W_i)$ holds that $\varphi_i(\bigcirc_\mu) \longmapsto \bigcirc_v$, then an arbitrary non-initial vertex of the complex U_i connected with $\bigcirc_v$ has the type of one of the vertexes of the subcomplex $L(W_i)$ connected with $\bigcirc_\mu$. Also, every complex U_i is a state of the class $\mathfrak{A}(\Gamma)$ of "conditions" satisfying the requirement $U(U_i) = U_i$. All $U_1,...,U_k$ are assumed to be non-isomorphic and W_i can be any state (i.e., either initial or terminal). Every pair of states is represented graphically in the following way. First, the complexes U_i are drawn on the left and the complexes W_i on the right. The boundary $L(U_i)$ is selected from the complex by framing it into two concentric rectangles. The same applies for the boundary $L'(W_i)$. The figure that follows depicts an isomorphic a pair of states:

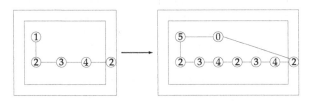

The isomorphism φ_i between $L(U_i)$ and $L'(W_i)$ is obtained by superimposing the right-hand part of the graphical representation to the left-hand part and by placing the coinciding vertexes between the rectangles.

The domain $\mathfrak{D}(\Gamma)$ of the operation $\Omega_\Gamma(S) = S'$ contains only those states S whose active part $U(S)$ is isomorphic to one of the complexes $U_1,...,U_k$.

Assume that the subcomplex $U(S)$ is isomorphic to U_i. Then, since $U(S)$ and U_i are connected complexes that belong to $\mathfrak{S}_{\mathfrak{Z}}$, if they are isomorphic, there is only one isomorphic correspondence between them, which uniquely determines an isomorphism between $L(S)$ and $L(U_i)$. Furthermore, since there is an isomorphic mapping φ_i from the complexes $L(U_i)$ to the complexes $L'(W_i)$, this induces a uniquely determined isomorphism θ_i^S between $L(S)$ and $L'(W_i)$.

Assume that $S \in \mathfrak{D}(\Gamma)$, where the subcomplex $U(S)$ is isomorphic to the complex U_i. Then, one can construct a complex W' that is isomorphic to the complex W_i and which satisfies the following conditions:

(i) $W' \cup V(S) = L(S)$; and

(ii) the isomorphism θ_i^S between the subcomplexes $L(S) \subseteq W'$ and $L'(W_i) \subseteq W_i$ can be extended to an isomorphism between the complexes W' and W_i.

The application of Ω_Γ to the complex S is uniquely, up to isomorphism, defined to be a complex of the form

$$S' = W' \cup V(S).$$

Suppose that S is a terminal state. Then, the connected component[5] of the initial vertex is considered to be the "solution." The set of connected components of the initial vertexes of terminal states is denoted by $\mathfrak{B}(\Gamma)$.

---

5. A connected component of any vertex is the maximal connected subcomplex that contains this vertex.

2.3.4   What Can Be Computed with These Algorithms?

Grigoriev [61] had argued that Kolmogorov–Uspensky algorithms can recognize a predicate that no machine with polynomial-limited accessible memory and in particular no *multidimensional* Turing machine can recognize.[6] Here memory is a storage structure that is defined in [36] as follows:

Definition 2.3.3 A storage structure $A = \langle L, \varphi_1, ..., \varphi_p \rangle$ of rank p consists of a countable set L (the locations or cells of Λ) together with the maps $\varphi_1,...,\varphi_p$ of L into L, called *shift transformations*.

For example, a multitape Turing machine has a storage structure of rank 2 and the two shift transformations are the left and the right shifts. Also, a polynomial-limited accessible memory is a storage structure defined as follows:

Definition 2.3.4 A storage structure or memory $A = \langle L, \varphi_1, ..., \varphi_p \rangle$ has polynomial-limited accessibility provided there are constants K and n such that the number of locations accessible in t steps from any given location does not exceed Kt^n. Here a location y is accessible from a location x in t steps provided there is a sequence $\psi_1, \psi_2,...,\psi_t$ of shift transformations such that $y = \psi_1 \psi_2 \cdots \psi_t(x)$.

In order to be able to fully grasp the ideas below, it is necessary to introduce some material from graph theory.

Preliminary Material from Graph Theory

A binary tree is called *directed* if there are either two edges or no edges coming from each node, and except for the root node, there is exactly one edge entering each node. Also, suppose that each edge of a binary tree is labeled by 0 or 1 so that two edges leaving the same node are labeled differently. A tree whose nodes, except the root, are labeled with 0 and 1 is called a *labeled tree*. For reasons of brevity, I will call *binary words* those words whose "letters" belong to the set $\{0, 1\}$.

Given a directed branch, there is a binary word that is formed as follows: starting from the edge that leaves the root node, the word consists of the labels of each edge. Clearly, the length of a binary word is equal to the number of edges that make up the branch. To each branch corresponds a unique binary word as far it regards the other binary words produced by branches of the same tree. Thus, it is possible to construct an inverse mapping on some set of binary words taking a binary word A to a branch α_A of the labeled tree starting at the root.

Assume that Γ is a labeled tree and that A is a binary word. Then, $D_\Gamma(A)$ is the binary word whose k-th letter is the label of the node which is the end of the k-th edge of the branch α_A. Note that $|D_\Gamma(A)| = |A|$.

A binary tree is called *regular* if all of its directed branches starting from the root and finishing at a leaf node (i.e., a node with no children) have the same length, which is called

---

6. A multidimensional Turing machine is one in which the tape can be either two or three dimensional. In the case of a two-dimensional machine, the transition function is of the form $\delta : (Q \setminus H) \times \Gamma \to Q \times \Gamma \times \{L, R, D, U, N\}$, where D and U specify movement of the scanning head up and down, respectively.

the depth of the tree. A regular tree Γ of depth $2n$ will be called semiuniform when for any binary words A_1, A_2, B of length n such that $A_1 \neq A_2$, the right half of the words $D_\Gamma(A_1 B)$ and $D_\Gamma(A_2 B)$ are distinct. Also, a regular tree Γ of depth 2^L will be called uniform if for all $n, k \in \mathbb{N}$ such that $(n + 1)2^{k+1} \leq 2^L$ any regular subtree of Γ of depth 2^{k+1} whose root is located at depth $n2^{k+1}$ is semiuniform.

Lemma 2.3.1 *There is an infinite sequence of uniform trees Γ_0, Γ_1,... of depth 2^0, 2^1,..., respectively, such that for $k \leq n$, Γ_k is an initial subtree of depth 2^k of Γ_n.*

Locally Complex Functions

A function g is *almost complex* if there are numbers ℓ_0 and θ ($0 \leq \theta < 1$) such that for any $n, k \in \mathbb{N}$ and for any words A and C in the alphabet Q with $|A| = n2^{k+1}$ and $\ell_0 \leq |C| = 2^k$, there is no equivalence class (see Sect. B.1) of the relation $\equiv_{A,C}$ that contains more than $\beta^{\theta 2^k}$ members, where β is the number of elements of Q. (The relation $\equiv_{A,C}$ is defined as follows: $X_1 \equiv_{A,C} X_2$ if $|X_1| = |X_2| = |C|$ such that $g(AX_1 c_1 ... c_i) = g(AX_2 c_1 ... c_i)$, $i = 1, 2, ...\ell$.)

A function f is *locally complex* if there is an almost complex function g and two sequences $(\alpha_n)_{n \in \mathbb{N}}$ and $(\beta_n)_{n \in \mathbb{N}}$ such that $\lim \sup_{k \to \infty} \beta_k = \infty$, where

$$\lim_{n \to \infty} \sup = \lim_{n \to \infty} \left(\sup_{m \geq n} \beta_m \right) = \inf \left\{ \sup\{\beta_m : m \geq n\} : n \geq 0 \right\},$$

and for all words A and B for which

$$|A| = \sum_{i=1}^{k} \alpha_i + \sum_{j=1}^{k-1} \beta_j \quad \text{and} \quad |B| \leq \beta_k$$

the equality $f(AB) = g(B)$ holds.

Lemma 2.3.2 *No locally complex function f is recognizable in real time by a system with polynomial-limited accessible memory.*

The Non-recognizable Predicate

The predicate that is not recognizable by a machine with polynomial-limited accessible memory, whereas it is computable by some Kolmogorov–Uspensky algorithm in real time, involves a locally complex function F that takes as arguments binary words and returns either the digit one or the digit zero.

Suppose that a fixed Kolmogorov–Uspensky algorithm K' constructs the tree Γ_L in T_L units of time, where the sequence $(T_L)_{L \in \mathbb{N}}$ is monotone increasing in L. Let us now define the values of function F. Given a binary word A, $F(A) = 1$ if

$$\sum_{i=0}^{L} T_i + 2^{L+1} \leq |A| < \sum_{i=0}^{L+1} T_i + 2^{L+1}$$

for some L; otherwise, if

$$\sum_{i=0}^{L-1} T_i + 2^{L-1} \le |A| < \sum_{i=0}^{L-1} T_i + 2^L$$

for some L, then $F(A)$ is the last letter in the word $D_{\Gamma(L)}(B)$ where

$$A = CB \quad \text{and} \quad |C| = \sum_{i=0}^{L-1} T_i + 2^{L-1} - 1$$

for some binary word C. The predicate that is not recognizable by a machine with polynomial-limited accessible memory, and, thus, by a Turing machine, is $P(A) \Leftrightarrow F(A) = 0$, where A is a binary word.

It is not difficult to devise a Kolmogorov–Uspensky algorithm K that recognizes P in real time. For each L, starting at time $\sum_{i=0}^{L-1} T_i + 2^i$ and finishing at time $\sum_{i=0}^{L} T_i + 2^i - 1$, algorithm K with the help of K$'$ builds the tree Γ_L. Next, if the input is a word B of length 2^L, the algorithm K does not build the tree Γ_L along the branch α_B and yields the word $D_{\Gamma_L}(B)$. In the time interval from $\sum_{i=0}^{L-1} T_i + 2^L$ to $\sum_{i=0}^{L} T_i + 2^L - 1$ the output of K is 1.

Lemma 2.3.3 *The function F is locally complex.*

Theorem 2.3.1 *There is a predicate not recognizable in real time by any system with polynomial-limited accessible memory but recognizable in real time by some Kolmogorov–Uspensky algorithm.*

This result can be proved using Lemmas 2.3.2 and 2.3.3 and algorithm K.

2.4 General Recursive Functions

In Sect. 2.1 it was explained how Turing machines compute functions and what makes a function computable by a Turing machine. Despite of this, a question that may naturally pop into one's mind is: Is it possible to tell whether a function is computable by a Turing machine without actually constructing the machine? There is class of numerical functions, that is, functions from nonnegative integers to nonnegative integers, that have exactly this property. These functions are called *recursive* functions. In fact, there are two "subclasses"—primitive and general recursive functions. In what follows, I will present these classes of functions. The exposition of the theory presented in this section is based on a seminal paper by Stephen Cole Kleene [73]. I will start with primitive recursive functions.

Primitive recursive functions are defined in terms of basic functions and function builders. There are three basic or initial functions:

(i) the *successor* function $S(x) = x + 1$;

(ii) the *zero* function $z(x) = 0$; and

(iii) the *projection* functions $U_i^n(x_1, ..., x_n) = x_i, 1 \le i \le n$.

Primitive recursive functions can be defined by applying function builders, or schemas, to the basic functions. There are three function builders:

Composition Suppose that f is a function of m arguments and each of $g_1, ..., g_m$ is a function of n arguments. Then the function obtained by composition of f and $g_1, ..., g_m$ is the function h defined as follows:

$$h(x_1, ..., x_n) = f\Big(g_1(x_1, ..., x_n), ..., g_m(x_1, ..., x_n)\Big).$$

Primitive Recursion A function h of $k + 1$ arguments is said to be definable by (primitive) recursion from the functions f and g, having k and $k + 2$ arguments, respectively, if it is defined as follows:

$$h(x_1, ..., x_k, 0) = f(x_1, ..., x_k),$$
$$h(x_1, ..., x_k, S(m)) = g\Big(x_1, ..., x_k, m, h(x_1, ..., x_k, m)\Big).$$

Minimalization The operation of minimalization associates with each total function f of $k + 1$ arguments a function h of k arguments. Given a tuple $(x_1, ..., x_k)$, the value of $h(x_1, ..., x_k)$ is the least value of x_{k+1}, if one exists, for which $f(x_1, ..., x_k, x_{k+1}) = 0$. If no such x_{k+1} exists, then its value is undefined.

Now we are ready to define primitive recursive and general recursive functions.

Definition 2.4.1 The functions that can be obtained from the basic functions by the function builders composition and primitive recursion are called primitive recursive functions.

Definition 2.4.2 The functions that can be obtained from the basic functions by all function builders are called general recursive functions.

Note that general recursive functions are also known as just recursive functions or μ-recursive functions.

We can easily extend the two previous definitions to define A-primitive recursive and A-recursive functions. However, in order to do this, we need to know what the characteristic function of a set is.

Definition 2.4.3 Assume that X is a universe set and $A \subseteq X$. Then the *characteristic function* $\chi_A : X \to \{0, 1\}$ of A is defined as follows:

$$\chi_A(a) = \begin{cases} 1, & \text{if } a \in A, \\ 0, & \text{if } a \notin A. \end{cases}$$

Note that this particular way of defining a set is actually employed to define fuzzy subsets, multisets, etc., via different types of characteristic functions, but I will say more on this in the next chapter. We are now prepared to define A-primitive recursive and A-recursive functions. Assume that $A \subseteq \mathbb{N}$ is a fixed set.

Definition 2.4.4 A function f is a partial A-recursive function if $f = \Psi_{\mathcal{M}}^A$, where $\Psi_{\mathcal{M}}^A$ is a partial function that denotes the computation performed by an oracle Turing machine $\mathcal{M}$ with oracle A.

Definition 2.4.5 A function f is an A-recursive function if there is an oracle machine $\mathcal{M}$ with oracle A such that $f = \Psi_{\mathcal{M}}^A$ and $\Psi_{\mathcal{M}}^A$ is a total function.

Definition 2.4.6 A set B is recursive in A if χ_B is A-recursive.

Kleen's s-m-n theorem, also called the iteration or the parametrization theorem, is a basic result in recursion theory. In what follows, I will present this important result without a proof.

Any Turing machine is characterized by a set of quadruples, where any two distinct quadruples must differ in their first or second part. This is known as the *consistency restriction*. Obviously, each quadruple is just a string of specific symbols. Furthermore, one can construct strings of these symbols only that, nevertheless, are not quadruples. Also, an algorithmic test would exist for determining, given such a string, whether it represents a quadruple or not. Hence, it is possible to list all sets of quadruples that satisfy the consistency restriction. In particular, such a procedure can arrange the list of sets of quadruples in such a way that each set of quadruples is associated with an integer x. More specifically, each integer x will be associated with the $(x + 1)$st element of the list. Each element of this list will be denoted by P_x. In addition, $\varphi_x^{(k)}$ is the partial function of k variables determined by P_x, where x is called an index of this function. Usually, the subscript is omitted when its value is clear from the context or when $k = 1$. Now, it is possible to state the s-m-n theorem:

Theorem 2.4.1 *For every $n, m \geq 1$, there exists a recursive function s_n^m of $m + 1$ variables such that for all $x, y_1, \ldots, y_m$,*

$$\lambda z_1 \cdots \lambda z_n . \varphi_x^{(m+n)}(y_1, \ldots, y_m, z_1, \ldots, z_n) = \varphi_{s_n^m(x, y_1, \ldots, y_m)}^{(n)}.$$

Given a partial recursive function φ_x, the symbol W_x will denote its domain.

2.5   Recursive Sets, Relations, and Predicates

It is quite natural to extend the notion of recursiveness to characterize not only functions but also sets, relations, and predicates. Informally, a set is called recursive if we have an effective method to determine whether a given element belongs to the set. However, if this effective method cannot be used to determine whether a given element does *not* belong to the set, then the set is called semirecursive. Formally, a recursive set is defined as follows.

Definition 2.5.1 Let $A \subseteq \mathbb{N}$ be a set. Then we say that A is primitive recursive or recursive if its characteristic function χ_A is primitive recursive or recursive, respectively.

Example 2.5.1 Suppose that Π is the set of all odd natural numbers. Then Π is primitive recursive, since its characteristic function

$$\chi_\Pi(a) = R(a, 2)$$

is primitive recursive. Here, $R(x, y)$ returns the remainder of the integer division $x \div y$.

Definition 2.5.2 A set A is called recursively enumerable or semirecursive either if $A = \varnothing$ or if A is the range of a recursive function.

The following result gives another characterization of recursively enumerable sets:

Theorem 2.5.1 *A set A is recursively enumerable if and only if A is the domain of a partial recursive function.*

A-recursively enumerable sets are defined as follows:

Definition 2.5.3 A set B is called A-recursively enumerable either if $B = \varnothing$ or if B is the range of an A-recursive function.

Definitions 2.5.1 and 2.5.2 can be easily extended to characterize *relations*. Note that an n-ary relation on a set A is any subset R of the n-fold Cartesian product $A \times \cdots \times A$ of n factors.

Definition 2.5.4 A relation $R \subseteq \mathbb{N}^m$ is called primitive recursive or recursive if its characteristic function χ_R given by

$$\chi_R(x_1, ..., x_m) = \begin{cases} 1, & \text{if } (x_1, ..., x_m) \in R, \\ 0, & \text{if } (x_1, ..., x_m) \notin R \end{cases}$$

is primitive recursive or recursive, respectively.

Definition 2.5.5 A relation $R \subseteq \mathbb{N}^m$ is called recursively enumerable (or semirecursive) if R is the range of a partial recursive function $f : \mathbb{N} \to \mathbb{N}^m$.

Let us now see how the notion of recursiveness has been extended to characterize predicates. But first let us recall what a predicate is. Roughly, it is a statement that asserts a proposition that must be either true (denoted by $t\!t$) or false (denoted by $f\!f$). An $\mathbb{N}^n$ function whose range of values consists exclusively of elements of the set $\{t\!t, f\!f\}$ is a predicate.

Definition 2.5.6 A predicate $P(x_1, ..., x_n)$ is called recursive if the set

$$\big\{(x_1, ..., x_n) \mid P(x_1, ..., x_n)\big\},$$

which is called its *extension*, is recursive.

Definition 2.5.7 The predicate $P(x_1, ..., x_n)$ is called recursively enumerable (or semirecursive) if there exists a partially recursive function whose domain is the set

$$\big\{(x_1, ..., x_n) \mid P(x_1, ..., x_n)\big\}.$$

The Arithmetic Hierarchy   Let us denote by Σ_0^0 the class of all recursive subsets of $\mathbb{N}$. For every $n \in \mathbb{N}$, Σ_{n+1}^0 is the class of sets that are A-recursively enumerable for some set $A \in \Sigma_n^0$. It follows that Σ_1^0 is the class of recursively enumerable sets. Let us denote by Π_0^0 the class of all subsets of $\mathbb{N}$ whose complements are in Σ_0^0. In other words, $D \in \Pi_0^0$ if and only if $\mathbb{N} \setminus D \in \Sigma_0^0$. The class Π_1^0 is known in the literature as the class of *co-recursively enumerable* sets. Any recursively enumerable set that is also co-recursively enumerable is a *decidable* set. In fact, the term decidable set is a synonym of the term recursive set.

Let us denote by Δ_n^0 the intersection of the classes Σ_n^0 and Π_n^0 (i.e., $\Delta_n^0 = \Sigma_n^0 \cap \Pi_n^0$). The classes Σ_n^0, Π_n^0, and Δ_n^0 form a hierarchy that is called the *arithmetic hierarchy*. The classes that make up this hierarchy have the following properties:

$$\Delta_n^0 \subset \Sigma_n^0, \qquad\qquad \Delta_n^0 \subset \Pi_n^0,$$
$$\Sigma_n^0 \subset \Sigma_{n+1}^0, \qquad\qquad \Pi_n^0 \subset \Pi_{n+1}^0,$$
$$\Sigma_n^0 \cup \Pi_n^0 \subset \Delta_{n+1}^0, \forall n \geq 1.$$

Figure 2.2 depicts the relationships between the various classes of the arithmetic hierarchy.

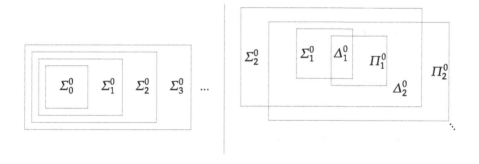

Figure 2.2: The relationships between the various classes of the arithmetic hierarchy

The set-theoretic presentation of the arithmetic hierarchy is not the only possible presentation. Indeed, other presentations based on predicates or relations are also possible. Assume that φ is a formula[7] in the language of first-order arithmetic (i.e., there are no other nonlogical symbols apart from constants denoting natural numbers, primitive recursive functions, and predicates that can be decided primitive recursively). Then, ϕ is a Δ_0^0-formula if ϕ contains at most bounded quantifiers. φ is Σ_1^0 if there is a Δ_0^0-formula $\psi(x)$ such that $\varphi \equiv (\exists x)\psi(x)$. Dually, φ is Π_1^0 if $\neg\varphi$ is Σ_1^0. More generally, a formula φ is in Σ_{n+1}^0 if there is a formula $\psi(x)$ in Π_n^0 such that $\varphi \equiv (\exists x)\psi(x)$. Dually, φ is Π_{n+1}^0 if $\neg\varphi$ is Σ_{n+1}^0. Note that Σ_0^0 is the class of all recursive predicates. Suppose that $Q_1 = P_1(x_1), Q_2 = P_2(x_1, x_2), Q_3 = P_3(x_1, x_2, x_3),\dots$ are Δ_0^0-formulas. Then Table 2.2 gives a schematic representation of the classes Σ_n^0 and Π_n^0.

---

7. Very roughly, a term yields a value; variables and constants are terms; functions are terms; atoms yield truth values; and each predicate is an atom. An atom is a formula. Given formulas p and q the following are also formulas: $\neg p$, $p \vee q$, $p \wedge q$, $p \Rightarrow q$, $p \equiv q$, $\forall x p$, and $\exists x p$. In the last two cases x is said to be a bound variable, while in all other possible cases it is said to be a free variable.

Table 2.2: A schematic representation of the classes Σ_n^0 and Π_n^0

	$n = 1$	$n = 2$	$n = 3$	
Σ_n^0	$(\exists x_1)Q_1$	$(\exists x_1)(\forall x_2)Q_2$	$(\exists x_1)(\forall x_2)(\exists x_3)Q_3$	...
Π_n^0	$(\forall x_1)Q_1$	$(\forall x_1)(\exists x_2)Q_2$	$(\forall x_1)(\exists x_2)(\forall x_3)Q_3$	...

Assume that $R \subseteq \mathbb{N}^m$ is a relation. Then $R \in \Sigma_1^0$ (i.e., R is Σ_1^0-relation) if R is recursively enumerable. Similarly, $R \in \Pi_1^0$ if $\overline{R} \in \Sigma_1^0$ (i.e., if the complement of R with respect to $\mathbb{N}^m$ is a Σ_1^0-relation). In general, $R \in \Sigma_n^0$ ($n \geq 2$) if there are a $k \in \mathbb{N}$ and a Π_{n-1}^0-relation $S \subset \mathbb{N}^{m+k}$ such that

$$R = \left\{ (x_1, ..., x_m) \mid \exists (x_{m+1}, ..., x_{m+k}) \in \mathbb{N}^k, (x_1, ..., x_{m+k}) \in S \right\}.$$

Also, $R \in \Pi_n^0$ if $\overline{R} \in \Sigma_n^0$.

Definition 2.5.8 A is Σ_n^0-complete if $A \in \Sigma_n^0$ and for all $B \in \Sigma_n^0$, $B \leq_1 A$, where $B \leq_1 A$ is pronounced B *is one–one reducible to* A and means that there is a one–one recursive function f such that for all $x \in B \Leftrightarrow f(x) \in A$.

Similarly, one can define Π_n^0-completeness.

Definition 2.5.9 A is *many–one reducible*, written as $A \leq_m B$, if there is a recursive function f such that for all x, $x \in A \Leftrightarrow f(x) \in B$.

Typically, the term "m-reducibility" is used instead of the term "many–one reducibility."

The Analytical Hierarchy   The second-order equivalent of the arithmetic hierarchy is called the *analytical hierarchy*. In this hierarchy, quantifiers range over function and relation symbols and over subsets of the universe. In other words, we are talking about second-order logic. A formula φ is a Π_1^1-formula if $\varphi \equiv (\forall X)\psi(X)$ and $\psi(X)$ is Σ_1^0. Dually, a formula φ is Σ_1^1 if and only if $\neg\varphi$ is Π_1^1. More generally, a formula φ is Π_{k+1}^1 if and only if $\varphi \equiv (\forall X)\psi(X)$ and $\psi(X)$ is Σ_k^1. Dually, φ is Σ_{k+1}^1 if and only if $\neg\varphi$ is Π_{k+1}^1. Clearly, it is easy to construct a table like Table 2.2 to provide a schematic representation of the analytical hierarchy. Note that the Δ_1^1-sets are the so-called *hyperarithmetic* sets. In addition, a function $f : \mathbb{N} \to \mathbb{N}$ is hyperarithmetic if its graph[8] G_f is a hyperarithmetic relation.

The arithmetic and analytic hierarchies are used to classify functions, sets, predicates, and relations. In particular, the higher the class an object belongs to, the more classically noncomputable it is.

2.6   The Church–Turing Thesis

In Sect. 2.4 I asked whether it is possible to tell if a function is computable by a Turing machine without actually constructing the machine. More generally, one can ask what functions

---

8. The graph of a function $f : X \to Y$ is the subset of $X \times Y$ given by $\{(x, f(x)) \mid x \in X\}$. A total function whose graph is recursively enumerable is a recursive function.

and/or numbers are computable by Turing machines? The cornerstone of classical computability theory and a direct answer to this question is the Church–Turing thesis, which can be phrased as follows.

Thesis 2.6.1 *Every effectively computable function is Turing computable, that is, there is a Turing machine that realizes it. Alternatively, the effectively computable functions can be identified with the recursive functions.*

Formally, a function $f : \mathbb{N}^n \to \mathbb{N}^m$ is Turing computable if there is a Turing machine $\mathcal{M}$ that computes it. But it is not clear at all what is meant by an *effective* procedure or method. Jack Copeland [37] gives four criteria that any sequence of instructions that make up a procedure or method should satisfy in order for it to be characterized as effective:

(i) each instruction is expressed by means of finite number of symbols;

(ii) the instructions produce the desired result in a finite number of steps;

(iii) they can be carried out by a human being unaided by any machinery save paper and pencil; and

(iv) they demand no insight or ingenuity on the part of the human carrying it out.

In his classical textbook [93], Marvin Minsky defines an effective procedure as "a set of rules which tell us, from moment to moment, precisely how to behave," provided we have at our disposal a universally accepted way to interpret these rules. Minsky concludes that this definition is meaningful if the steps are actually steps performed by some Turing machine. Another formulation of effectiveness is given in [66]:

> [E]very instruction must be sufficiently basic that it can in principle be carried out by a person using only pencil and paper. It is not enough that each operation be clear and unambiguous, but it must also be feasible.

Gábor Etesi and István Németi [48] describe as effectively computable any function $f : \mathbb{N}^k \to \mathbb{N}^m$ for which there is a *physical computer* realizing it. Here, by "realization by a physical computer" they mean the following:

> Let P by a physical computer, and $f : \mathbb{N}^k \to \mathbb{N}^m$ a (mathematical) function. Then we say that P *realizes* f if an imaginary observer O can do the following with P. Assume that O can "start" the computer P with $(x_1, \dots, x_k)$ as an input, and then sometime later (according to O's internal clock) O "receives" data $(y_1, \dots, y_m) \in \mathbb{N}^m$ from P as an output such that $(y_1, \dots, y_m)$ coincides with the value $f(x_1, \dots, x_k)$ of the function f at input $(x_1, \dots, x_k)$.

The same authors, after introducing the notion of *artificial computing systems*, that is, thought experiments relative to a fixed physical theory that involve computing devices, managed to rephrase the Church–Turing thesis as follows.[9]

---

9. Actually, they call this "updated" version of the Church–Turing thesis the *Church–Kalmár–Turing* thesis, named after Church, László Kalmár, and Turing.

Thesis 2.6.2 *Every function realizable by an artificial computing system is Turing computable.*

Since artificial computing systems are thought experiments relative to a fixed physical theory, the thesis can be rephrased as follows.

Thesis 2.6.3 *Every function realizable by a thought experiment is Turing computable.*

Note that according to Etesi and Németi, a thought experiment relative to a fixed physical theory is a theoretically possible experiment, that is, an experiment that can be carefully designed, specified, etc., according to the rules of the physical theory, but for which we might not currently have the necessary resources.

Others, like David Deutsch [42], have reformulated the Church–Turing thesis as follows.

Thesis 2.6.4 *Every finitely realizable physical system can be perfectly simulated by a universal model computing machine operating by finite means.*

In the special case of the human mind, this thesis can be rephrased as follows.

Thesis 2.6.5 *The human brain realizes only Turing-computable functions.*

This thesis is the core of computationalism. This philosophy claims that a person's mind is actually a Turing machine. Consequently, one may go a step ahead and argue that since a person's mind is a Turing machine, then it will be possible one day to construct an artificial person with feelings and emotions. The mind is indeed a machine, but one that transcends the capabilities of the Turing machine and operates in a profoundly different way (see [126] for more on this matter).

2.7   Computational Complexity

Roughly, there are two kinds of solvable problems that are not too difficult—those that can be solved easily and those that can be solved with difficulty. In particular, when a problem is characterized as difficult, then

- if there is an algorithmic solution to this problem, it can be checked quickly; and

- an algorithmic solution will require an impossibly long time to yield an output.

In different words, it is necessary to examine the operation of the Turing machine that solves a particular problem in order to decide whether a problem is difficult or not. Formally,

Definition 2.7.1 A Turing machine is *polynomial bounded* when there is polynomial $p(n)$ such that for any input x there is no configuration C such that

$$q \triangleright \sqcup x \vdash_{\mathscr{M}}^{p(|x|)+1} C,$$

where $p(|x|) + 1$ is the length of the computation. In different words, the machine always halts after at most $p(|x|)$ steps.

Definition 2.7.2 A language is called *polynomial decidable* if there is polynomially bounded Turing machine that decides it. The class of all polynomial decidable languages is denoted P.

Definition 2.7.3 A nondeterministic Turing machine is said to be *polynomial bounded* when there is a polynomial $p(n)$ such that for any input x, there is no configuration C such that

$$q \triangleright \llcorner x \vdash_{\mathcal{N}}^{p(|x|)+1} C.$$

In different words, no computation of this machine requires more than polynomially many steps.

Definition 2.7.4 The class of all languages that are decided by a polynomially bounded non-deterministic Turing machine is denoted NP.

3. Elements of Fuzzy Set Theory

The notion of fuzziness lies at the core of fuzzy computability theory. Thus, one should have a basic understanding of the ideas involved. This chapter serves both as a crash course in fuzzy set theory, for those readers that have no previous knowledge of the concepts involved, and as a précis of fuzzy set theory, for those readers familiar with the relevant notions. The exposition that follows is based on [75], while the material for Sect. 3.1 is borrowed from [49, 137].

3.1 Ordered Sets

Assume that P is a set that is equipped with a binary relation $\leqslant$ (i.e., a subset of $P \times P$). Then, if this relation satisfies the following laws:

reflexivity $a \leqslant a$ of all $a \in P$;

transitivity if $a \leqslant b$ and $b \leqslant c$, then $a \leqslant c$, for all $a, b, c \in P$; and

antisymmetry if $a \leqslant b$ and $b \leqslant a$, then $a = b$ for all $a, b \in P$;

the set P is called a *poset* or *partially ordered set*. Each poset has some special elements.

Definition 3.1.1 Suppose that P is a poset, $X \subseteq P$, and $y \in P$. Then, y is the *greatest lower bound* or *infimum* or *meet* for X if and only if

- y is a *lower bound* for X, that is, if $x \in X$, then $y \leqslant x$; and

- if z is any other lower bound for X, then $z \leqslant y$.

Typically, one writes $y = \bigwedge X$. If $X = \{a, b\}$, then the meet for X is denoted as $a \wedge b$.

Definition 3.1.2 Suppose that P is a poset, $X \subseteq P$, and $y \in P$. Then, y is the *least upper bound* or *supremum* or *join* for X if and only if

- y is an *upper bound* for X, that is, if $x \in X$, then $x \leqslant y$; and

- if z is any other upper bound for X, then $y \leqslant z$.

Typically, one writes $y = \bigvee X$. If $X = \{a, b\}$, then the join for X is denoted as $a \vee b$.

A. Syropoulos, *Theory of Fuzzy Computation*, IFSR International Series on Systems Science and Engineering 31, DOI 10.1007/978-1-4614-8379-3_3,
© Springer Science+Business Media New York 2014

Definition 3.1.3 A subset P' of a poset $(P, \preccurlyeq)$ is called *bounded above* if there is an element $a \in P$ such that $b \preccurlyeq a$ for all $b \in P'$. Similarly, one can define the notion of a *bounded below* subset of a poset. A subset P' of a poset $(P, \preccurlyeq)$ is called *bounded* when it is both bounded above and below.

Definition 3.1.4 Suppose that $(P, \preccurlyeq)$ is a poset. Then, a subset P' of P is called a *chain* in P if any two elements of P' are comparable. In addition, a chain is *complete* when every non-empty bounded subset has a meet and a join.

Definition 3.1.5 A subset P' of a partially ordered set $(P, \preccurlyeq)$ is called a *dense* (or, a *co-initial*) subset of P if and only if for every nonzero element a of P there exists a nonzero element b of P' such that $b \preccurlyeq a$.

Directed sets and lattices are special kinds of posets:

Definition 3.1.6 A *directed set* is a poset $(P, \preccurlyeq)$ such that whenever $a, b \in P$, there exists an $x \in P$ such that $a \preccurlyeq x$ and $b \preccurlyeq x$.

A lattice is a special kind of a poset:

Definition 3.1.7 A poset $(P, \preccurlyeq)$ is a *lattice* if and only if every finite subset has both a greatest lower bound and a least upper bound.

Assume $(L, \preccurlyeq)$ is a lattice and that $a, b \in L$. Then, a is said to be *way below* b, written $a \ll b$, if for any directed set $D \subseteq L$ such that $\bigvee D$ exists and that $b \preccurlyeq \bigvee D$, there is a $d \in D$ such that $a \preccurlyeq d$. Note that if $a \ll b$, then $a \leq b$ since one can set $D = \{b\}$. Also, if L is finite, the converse holds true.

A frame is also a special form of a poset:

Definition 3.1.8 A poset $(P, \preccurlyeq)$ is a *frame* if and only if

   (i) every subset has a least upper bound;

  (ii) every finite subset has a greatest lower bound; and

 (iii) binary greatest lower bounds distribute over least upper bounds:

$$x \wedge \bigvee Y = \bigvee \left\{ x \wedge y \mid y \in Y \right\}.$$

If the third clause of this definition is not valid, the resulting structure is a *complete* lattice.

Definition 3.1.9 A *based continuous lattice*, or just *based* lattice, is a structure $(L, \preccurlyeq, B)$, where L is a complete lattice and B, which is called the *basis*, is a subset of L that includes 0, closed under the meet and join operations and such that, for all $a \in L$,

$$a = \bigvee \left\{ b \ll a \mid b \in B \right\}.$$

When L is a finite chain and $B = L$, then for the resulting structure it holds that $b \ll a \Leftrightarrow b \leqslant a$. Also, if L is a complete chain and B is a dense subset of L, then $(L, \leqslant, B)$ is based lattice such that $b \ll a$ if and only if either $a = 0$ or $a \leqslant b$.

Theorem 3.1.1 *Assume that $(L, \leqslant, B)$ is a based lattice and that $a \ll b$. Then, there is a $x \in B$ such that $a \ll x \ll b$. In addition, for any directed family $(a_i)_{i \in I}$, $b \leqslant \bigvee_{i \in I} a_i$ implies that there is a a_i such that $a \ll a_i$.*

Definition 3.1.10 An *effective continuous* lattice (see [120]), or simply an *effective* lattice, is a based lattice $(L, \leqslant, B)$ with a sequence $(b_n)_{n \in \mathbb{N}}$ such that

- the relation $\left\{ (n, m) \mid \left((n, m) \in \mathbb{N} \times \mathbb{N} \right) \wedge \left(b_n \ll b_m \right) \right\}$ is recursively enumerable; and

- there exist two recursive functions join : $\mathbb{N} \times \mathbb{N} \to \mathbb{N}$ and meet : $\mathbb{N} \times \mathbb{N} \to \mathbb{N}$ with the following properties:

$$b_n \vee b_m = b_{\text{join}(n,m)} \text{ and } b_n \wedge b_m = b_{\text{meet}(n,m)}.$$

3.2   Fuzzy Subsets

Although readers have already been exposed to the notation that follows, still it makes no harm to summarize how sets can be defined. In particular, there are three methods to define a set:

List method   A set is defined by naming all its members. This method can be used only for finite sets. For example, any set A, whose members are the elements $a_1, a_2, ..., a_n$, where n is small enough, is usually written as

$$A = \{a_1, a_2, ..., a_n\}.$$

Rule Method   A set is defined by specifying a property that is satisfied by all its members. A common notation expressing this method is

$$A = \left\{ x \mid P(x) \right\},$$

where the symbol $\mid$ denotes the phrase "such that" and $P(x)$ designates a proposition of the form "x has the property P."

Characteristic function   This method was introduced in Definition 2.4.3.

All known extensions of the notion of a set are introduced by using one or more of these methods.

As was mentioned in the introduction, fuzzy sets are an extension of classical sets where elements are allowed to belong to a fuzzy subset to a degree. Typically, fuzzy subsets are defined using an extension of the characteristic function:

Definition 3.2.1 Given a universe X a fuzzy subset A of X is *characterized* by a function $A : X \to [0, 1]$ and $A(x)$ denotes the degree to which element x belongs to the fuzzy subset A.

Note that I have used the word *characterized* and not *is* since fuzzy subsets are an extension of classical mathematics and the function, that is, a classical structure, that characterizes the fuzzy subsets is defined as a mapping between classical sets. Also, it is common in the literature to talk about fuzzy sets when, in fact, the term refers to fuzzy subsets. Only for reasons of brevity I will sometimes follow this practice for the rest of this book. Let me now provide the definition of some common set operations.

Definition 3.2.2 Assume that $A : X \rightarrow [0,1]$ and $B : X \rightarrow [0,1]$ characterize two fuzzy subsets of X. Then

- their *union* is
$$(A \cup B)(x) = \max\{A(x), B(x)\};$$

- their *intersection* is
$$(A \cap B)(x) = \min\{A(x), B(x)\};$$

- the *complement* of A is the fuzzy subset
$$\bar{A}(x) = 1 - A(x) \quad \text{for} \quad x \in X;$$

- A is a *subset* of B, denoted by $A \subseteq B$, if and only if
$$A(x) \leq B(X) \, \forall x \in X; \text{ and}$$

- the *scalar* cardinality of A is
$$|A| = \sum_{x \in X} A(X).$$

As was explained in Sect. 3.1, the symbols $\wedge$ and $\vee$ are used to denote the infimum and the supremum of a poset. Now, it is customary, instead of the "words" min and max, to use the symbols $\wedge$ and $\vee$, respectively. This is justified by the fact that the suprema and the infima of the unit interval or any closed subsets of it, when seen as a poset, can be computed with the maximum and the minimum operations, respectively. Thus, I will follow this practice in the rest of this book; nevertheless, when confusion may arise, I will explicitly state what is really meant.

Two very important concepts of fuzzy set theory are the concept of an *α-cut* and of a *strong α-cut*:

Definition 3.2.3 Suppose that $A : X \rightarrow [0,1]$ characterizes a fuzzy subset of X. Then, for any $\alpha \in [0,1]$, the α-cut $^{\alpha}A$ and the strong α-cut $^{\alpha+}A$ are the ordinary sets

$$^{\alpha}A = \left\{x \mid (x \in X) \wedge \left(A(x) \geq \alpha\right)\right\}$$

and

$$^{\alpha+}A = \left\{x \mid (x \in X) \wedge \left(A(x) > \alpha\right)\right\},$$

respectively.

SpringerBriefs in Statistics

For further volumes:
http://www.springer.com/series/8921

Daniele Manfredini • Rosa Arboretti • Luca Guarda-
Nardini • Eleonora Carrozzo • Luigi Salmaso

Statistical Approaches to Orofacial Pain and Temporomandibular Disorders Research

 Springer

Daniele Manfredini
University of Padova
Padova
Italy

Rosa Arboretti
Dept. Land, Environment, Agriculture
and Forestry
University of Padova
Padova
Italy

Luca Guarda-Nardini
University of Padova
Padova
Italy

Eleonora Carrozzo
Department of Management and
Engineering
University of Padova
Padova
Italy

Luigi Salmaso
Department of Management and
Engineering
University of Padova
Padova
Italy

ISSN 2191-544X   ISSN 2191-5458 (electronic)
ISBN 978-1-4939-0875-2   ISBN 978-1-4939-0876-9 (eBook)
DOI 10.1007/978-1-4939-0876-9
Springer New York Heidelberg Dordrecht London

Library of Congress Control Number: 2014936406

Printed on acid-free paper

Springer is part of Springer Science+Business Media (www.springer.com)

Contents

Chapter 1
Fundamentals

Daniele Manfredini, Rosa Arboretti, Eleonora Carrozzo, and Luca Guarda-Nardini

This chapter, introductory in nature, is likely the core part of the book. It summarizes the available knowledge on temporomandibular disorders (TMD), a heterogeneous group of orofacial pain conditions that are the main subject of most investigations described and commented throughout the book. TMD are gaining attention from several medical professions, with more than 15,000 citations listed in the medline database at the end of the year 2013. This specific field of expertise is going through an epochal change because, after decades of mechanicistic approaches to their diagnosis and treatment, which were based on the correction of dental occlusion abnormalities, there is now consensus that TMD are musculoskeletal disorders requiring a multidimensional approach. Consequently, such a very complex field may reflect in several different types of investigations depending on the aims that are pursued. Also, the validity of findings is conditioned by the validity of the statistical design. In particular, several strategies can be identified to perform studies on the etiology, diagnosis, and treatment of the disease. In this chapter, a brief introduction to the importance of choosing the right statistical approach in the design of dental research is followed by the description of the current concepts on TMD clinics. This may serve as a fundamental basis for the readers to get through the different example investigations strategies that are described in the following chapters with more specific contents.

D. Manfredini (✉) · R. Arboretti · E. Carrozzo · L. Guarda-Nardini
University of Padova, Padova, Italy
e-mail: daniele.manfredini@tin.it

R. Arboretti
e-mail: rosa.arboretti@unipd.it

E. Carrozzo
e-mail: carrozzo@gest.unipd.it

L. Guarda-Nardini
e-mail: luca.guarda@unipd.it

L. Salmaso et al., *Statistical Approaches to Orofacial Pain and Temporomandibular Disorders Research*, SpringerBriefs in Statistics, DOI 10.1007/978-1-4939-0876-9_1, © The Author(s) 2014

1.1   The Importance of Statistics in Dentistry

Statistics is defined as the theory and methodology for study design and for describing, analyzing, and interpreting the data generated from such studies. Medical statistics deals with applications of statistics to medicine and the health sciences, including epidemiology, public health, forensic medicine, and clinical research. Good doctors have to understand and study statistics for several reasons, viz., updating their medical knowledge in the clinical setting, knowing basic requirements for medical research, managing and treating data in the research setting, and applying the right evidence-supported concepts in the clinical setting. The application of the right principles of statistics in medicine is becoming more and more important also because of the increasing social implications of the peer-reviewed publications, which influence allocation of resources and funding.

A growing attention to statistical issues characterized the recent literature of several medical fields, and dentistry was not an exception. Some publications and textbooks are available for those practitioners willing to get deeper into the arena of statistics in dentistry, but very few information focusing on orofacial pain patients is available. Within the dental profession, orofacial pain is a field requiring a peculiar knowledge in the clinical setting, where combined expertise in different medical branches is strongly requested for managing the difficult pathway to differential diagnosis, as well as in the research setting, where the choice of the right study design is often complicated by the epidemiological features of the disease.

TMD are a subset of orofacial pain disorders gaining attention from several medical professions. In such a very complex field, different types of investigations can be performed depending on the aims that are pursued, and the validity of findings is conditioned by the validity of the statistical design. In particular, several strategies can be identified to perform studies on the etiology, diagnosis, and treatment of the disease. For instance, at the etiological level, it is fundamental that the identification of risk factors for TMD is based on multiple variable analysis taking into account for the complexity of the biological model in which disease can occur. Also, a study assessing the diagnostic accuracy of an instrument needs to test such an accuracy in terms of the instrument's capability to distinguish patients from non-patients, viz., subjects with pain versus those without pain. Pain assessment, possibly in a multimodal setting, is also the target of all studies on treatment effectiveness.

For a better comprehension of the statistical designs, some details on the current concepts on TMD are worthy to be summarized.

1.2   TMD: Definition of the Problems in Clinical Research

TMD are a heterogeneous group of pathologies affecting the temporomandibular joint (TMJ), the masticator muscles, or both (Okeson 2008). TMD may present with a number of signs and symptoms, the most common of which are pain localized in the preauricular area and/or in the masticatory muscles; jaw motion abnormalities;

and articular sounds, such as click and/or crepitus, during mandibular movements (Laskin 1969). A specific etiopathogenesis is rarely demonstrable, since most cases seem to have a multifactorial etiopathogenetic pathway (Greene 2006a). Epidemiological data showed a female predominance, which is more marked in patients' populations, and a mean age of onset around 35–45 years, with two distinct age peaks for internal joint derangements and inflammatory-degenerative disorders (Schiffmann et al. 1990; Manfredini et al. 2006, 2010).

The complex etiopathogenesis and the variability of symptoms make difficult to adopt standardized diagnostic and therapeutic protocols, thus reflecting in the proposal of several different treatment approaches, such as occlusal splints (Klasser and Greene 2009a), physiotherapy (Gavish et al. 2006), behavioral treatments (Gatchel et al. 2006), physical therapy (De Laat et al. 2003), drugs (Dionne 2000), and minor (Nitzan et al. 1991; Guarda-Nardini et al. 2007) and major surgery (Dolwick and Dimitroulis 1994; Guarda-Nardini et al. 2008).

In the recent years, many progresses have been made in the attempt to design reference principles for the diagnosis and treatment. This led to the diffusion of internationally recognized academic guidelines for the assessment and management of TMD patients in the clinical setting (De Boever et al. 2008; De Leeuw 2008b), to the adoption of a standardized protocol translated in several languages, viz., the Research Diagnostic Criteria for TMD (Dworkin and Leresche 1992), for the diagnosis and classification of such disorders in the research setting, and to a recent updated classification aiming to reach both targets (Schiffman et al. 2014). Also, some seminal articles provided evidence-based invoices for the adoption of low-technology, high-prudence, conservative, and reversible approaches to TMD (Stohler and Zarb 1999; Manfredini et al. 2012a). Nevertheless, non-specialist and non-expert practitioners still refer many uncertainties at both diagnostic and therapeutic levels, thus suggesting that the quality of communication between the research and clinical settings, viz., the science transfer process, should be enhanced (Greene 2006b; Manfredini 2010b). In particular, it seems that the well-documented view of TMD as non-dentally related disorders (Koh and Robinson 2003; Turp et al. 2008) is hard to be accepted by the general dental practitioners, who had been accustomed for years to provide occlusally based treatments to their TMD patients and are reluctant to accept any paradigmatic shifts in their daily practice (Klasser and Greene 2009b; Turp and Schindler 2010).

The problems of diffusion of evidence-based knowledge into the clinical TMD setting might influence several issues concerning the TMD practice, so that groups of front-line experts, researchers, and academicians are strongly encouraged to educate themselves to the methodological and statistical issues concerning the research design.

To do that, the reasons for the discrepancies between the suggestions coming from the literature and the unsupported ideas that are still diffused among some clinicians need to be analyzed.

Three levels of difficulty in the process of transferring scientific information from the research setting to the clinical practice can be identified: (1) methodological difficulties to conduct researches on TMD patients; (2) difficulties in the researchers-to-clinicians communication phase; (3) difficulties for clinicians to accept ideas that might change their practice.

The main problems that researchers have to face when designing an investigation on TMD patients are the absence of a known physiopathology and the difficulty to define "success" at the treatment level. The definition of "success" when treating patients with orofacial pain conditions such as TMD is strictly related to the knowledge about the disease's etiology and physiopathology. The limited knowledge about the TMD etiology prevents from pursuing a causative therapy and, consequently, to define a successful treatment on the basis of the eradication of the causal factor. There is now consensus that TMD treatment must be based on symptoms' management (Manfredini 2010a). This represents the best choice in the clinical setting, but it brings out some problems if one attempts to transfer such a concept in the research setting.

The main question researchers have to answer in the design phase of an investigation is: Which improvement in which outcome variable is clinically significant?

Answering to this question is a compulsory need before conducting a power analysis and establishing the needed sample size for a study. At present, very few studies have addressed the issue of power analysis and sample size calculation for TMD studies (Dao et al. 1994; Manfredini et al. 2003). The difficulties that prevent from achieving a standardized power analysis are different for studies dealing with the etiological, diagnostic, or therapeutic aspects. In particular, studies assessing either the accuracy of a technique to diagnose TMD or the efficacy of a treatment should be based on pain rating as the main outcome variable, but the assessment of TMD-related pain is complicated by the high rate of psychosocial impairment that typically characterizes TMD patients populations (Rollmann and Gillespie 2000). This reduces the possibility to provide an objective and standardized evaluation of pain. An assessment of TMD symptoms is also complicated by the self-limiting and fluctuating nature of the disease. Indeed, symptoms' fluctuation has been reported both in the short- and long-term periods (Magnusson et al. 2005), and several studies have described a tendency toward a spontaneous remission in the majority of cases (De Leeuw 2008a).

These observations might have confused the issue and authorized some clinicians to treat "non-patients" and feel that their treatment was effective. Obviously, this is an extreme simplification of the relation between TMD research and practice, but it seems clear that a strong emphasis should be put on the need for a thorough and standardized assessment. Patients populations would be defined better, and "non-patients" would be excluded from the investigations in the research setting and would avoid unnecessary treatments in the clinical setting.

Moreover, once established standardized criteria to compare study populations, information on the relative efficacy of the different treatment approaches should be more consistent, and treatments should be compared in terms of cost effectiveness and risk to benefit ratio.

1.3   The Problem of Sample Size

The aim of most researches is to investigate for differences between the value of a parameter in a study population and a hypothesized value (the hypothesized value is usually that of the same parameter in a reference population). If the p value is

A special kind of α-cut is the *support* of a fuzzy set A. In particular, the support of A, denoted by $S(A)$, is the set of all elements of X that have nonzero membership degrees, that is, the strong α-cut ^{0+}A.

Remark 3.2.1 A fuzzy set $A : X \rightarrow [0,1]$ is characterized by a subset of the Cartesian product $S(A) \times [0,1]$.

3.3   L-Fuzzy Subsets

It does make sense to relax the requirement that the membership degrees are drawn from the unit interval. More generally, one can assume that the membership degrees are drawn from sets that are at least partially ordered. Typically, these new types of sets are called L-fuzzy sets and were introduced by Joseph Amadee Goguen [54]. However, it is customary to call L-fuzzy sets those whose membership degrees are drawn from a lattice or a frame, in some cases. The following defines the basic set operations:

Definition 3.3.1 Assume that $A : X \rightarrow L$ and $B : X \rightarrow L$ characterize two L-fuzzy subsets of X. Then,

- their *union* is
$$(A \cup B)(x) = A(x) \vee B(x);$$

- their *intersection* is
$$(A \cap B)(x) = A(x) \wedge B(x); \ and$$

- the *complement* of A is the L-fuzzy subset
$$\bar{A}(x) \ such \ that \ \bar{A} \vee A(x) = 1 \ and \ \bar{A} \wedge A(x) = 0, \ \forall x \in X,$$

where
$$\bigvee \emptyset = 0 \quad and \quad \bigwedge \emptyset = 1$$
are the bottom and the top elements of the lattice.

3.4   Fuzzy Relations

Definition 3.4.1 Given two sets A and B a binary fuzzy relation R between these sets is characterized by a function $R : A \times B \rightarrow [0,1]$.

Typically, a fuzzy relation between two sets A and B is simply written as $R(A, B)$.

Definition 3.4.2 The *complement* of a fuzzy relation $R(A, B)$ is a fuzzy relation $\bar{R}(A, B)$ such that $R(x, y) = 1 - \bar{R}(x, y)$.

Definition 3.4.3 Assume that $P(X, Y)$ and $Q(Y, Z)$ are two binary fuzzy relations with a common set Y. Then, the *composition* of these relation, which is denoted by $P(X, Y) \circ Q(Y, Z)$, is a binary fuzzy relation $R(X, Z)$ defined by

$$R(x, z) = \max_{y \in Y} \min \left[P(x, y), Q(y, z) \right],$$

for all $x \in X$ and all $z \in Z$. The composition just defined is called the standard composition, but one can define others using t-norms and t-conorms (see next section).

Given a fuzzy relation $R(A, A)$, its *transitive closure* $R_T(A, A)$ is a fuzzy relation that is transitive and contains $R(A, A)$, and its elements have the smallest possible membership degrees. The transitive closure of a fuzzy binary relation $R(A, A)$ can be computed by the following simple steps:

(i) $R' = R \cup (R \circ R)$;

(ii) if $R' \neq R$, then $R \leftarrow R'$ and goto step 1; and

(iii) $R_T \leftarrow R'$.

Assume that $\mathbb{N} \ni p \neq 0$ and X is a universe. Then, a fuzzy subset $R : X^p \rightarrow [0, 1]$ is a *p-ary fuzzy relation on* X. The *projection* of a $p + 1$-ary fuzzy relation $R' : X^{p+1} \rightarrow [0, 1]$ is the p-ary relation R defined as follows:

$$R(x_1, x_2, ..., x_p) = \bigvee \left\{ R'(x_1, ..., x_p, x_{p+1}) \mid x_{p+1} \in X \right\}.$$

The *co-projection* of R' is defined as follows:

$$R(x_1, x_2, ..., x_p) = \bigwedge \left\{ R'(x_1, ..., x_p, x_{p+1}) \mid x_{p+1} \in X \right\}.$$

3.5   Triangular Norms and Conorms

Triangular norms or, usually, just t-norms are binary operations that generalize intersection in a lattice and conjunction in logic.

Definition 3.5.1 A t-norm is a binary operation $* : [0, 1] \times [0, 1] \rightarrow [0, 1]$ that satisfies at least the following conditions for all $a, b, c \in [0, 1]$:

boundary condition $a * 1 = a$ and $a * 0 = 0$;

monotonicity $b \leq c$ implies $a * b \leq a * c$;

commutativity $a * b = b * a$; and

associativity $a * (b * c) = (a * b) * c$.

The following are examples of some t-norms that are frequently used in practice:

standard intersection $a * b = \min(a, b)$;

algebraic product $a * b = ab$;

bounded difference $a * b = \max(0, a + b - 1)$; and

drastic intersection $a * b = \begin{cases} a, & \text{when } b = 1, \\ b, & \text{when } a = 1, \\ 0, & \text{otherwise.} \end{cases}$

Dually, t-conorms are binary operations that generalize union in a lattice and disjunction in logic.

Definition 3.5.2 A t-conorm is a binary operation $\star : [0, 1] \times [0, 1] \to [0, 1]$ that satisfies at least the following conditions for all $a, b, c \in [0, 1]$:

boundary condition $a \star 0 = a$ and $a \star 1 = 1$;

monotonicity $b \leq c$ implies $a \star b \leq a \star c$;

commutativity $a \star b = b \star a$; and

associativity $a \star (b \star c) = (a \star b) \star c$.

The following are examples of some t-conorms that are frequently used in practice:

standard union $a \star b = \max(a, b)$;

algebraic sum $a \star b = a + b - ab$;

bounded sum $a \star b = \min(1, a + b)$; and

drastic union $a \star b = \begin{cases} a, & \text{when } b = 0, \\ b, & \text{when } a = 0, \\ 1, & \text{otherwise.} \end{cases}$

3.6   Some Thoughts on Fuzzy Set Theory

The novelty introduced by fuzzy set theory is that sets contain elements to some degree or that elements belong to sets to some degree. But if we are not absolutely sure whether some element belongs to a set, how can we be sure whether two sets, or, for that matter, two elements, are surely equal? Furthermore, does it make sense to say that some fuzzy subset is a subset of another fuzzy subset? Unfortunately, these questions have not been addressed in the mainstream fuzzy theoretic literature. In what follows, I will briefly discuss an "extension" to fuzzy set theory that addresses these problems... to a certain degree.

Kosko [80] has define a fuzzy notion of subsethood, but it seems that since set theory is built on membership and equality, it is almost useless to fuzzify subsethood without fuzzifying equality. Nevertheless, his subsethood theorem is an attempt to provide an estimation of the degree to which one fuzzy subset is a subset of another one.

Theorem 3.6.1 *Assume that X is a universe and A and B two fuzzy subsets of X. Then, the degree to which A is a subset of B, denoted by S(A, B), is*

$$S(A, B) = \frac{|A \cap B|}{|A|}.$$

Michael Barr [6] has given a solution to the fuzzification of equality. In particular, given some universe X, a fuzzy set is now a pair ($\xi : X \to [0, 1], \epsilon : E \to [0, 1]$) such that E is an *equivalence relation* on X.[1] In general, if $(x, y) \in E$ with $\epsilon(x, y) = t$ while $\xi(x) = r$ and $\xi(y) = s$, then $t \leq r$ and $t \leq s$. This means that x and y belong to the fuzzy subset with degrees that are equal to r and s, respectively, while $x = y$ to degree that is equal to t. Using this idea it is possible to define a fuzzier theory of fuzzy sets in which both the membership and the equality are fuzzy. However, Barr proposes a "better" formulation of a fuzzier fuzzy set theory that is defined in the language of category theory. This alternative formulation leads to the definition of a category that is a topos (i.e., a special category that is a kind of intuitionistic mathematical universe), but a proper presentation of these ideas falls outside the scope of this book.

---

1. Recall that an equivalence relation R on a set X is a subset of $X \times X$ (i.e., $R \subseteq X \times X$) such that $a\,R\,a$ for all $a \in X$, $a\,R\,b$ implies $b\,R\,a$ for all $a, b \in X$, and $a\,R\,b$ and $b\,R\,c$ imply $a\,R\,c$ for all $a, b, c \in X$.

4. On Fuzzy Turing Machines

The Turing machine is the archetypal conceptual computing device and as such it is almost always used to define new models of computations by augmenting the functionality of the machine. For example, as was explained in Sect. 2.2.2, a probabilistic model of computation is defined by introducing a probabilistic version of the Turing machine, that is, a machine that operates nondeterministically. Therefore, it was quite expected to see the emergence of fuzzy Turing machines as models of fuzzy computation, that is, computation that encompasses vagueness in the form of fuzziness as a basic ingredient of computation.

4.1 Early Ideas

As expected, the first researcher who intuitively discussed the idea of *fuzzy algorithms* and, consequently, that of fuzzy Turing machines was Zadeh [145] himself. According to Zadeh, a fuzzy algorithm may contain fuzzy commands. For example, the following command is what Zadeh meant when he wrote about fuzzy commands:

> Make y approximately equal to 10, if x is approximately equal to 5.

The reason why this command is fuzzy is because it makes use of fuzzy sets. More generally, one could argue that a number of everyday activities can be viewed as fuzzy algorithms. For instance, the directions for repairing a computer or constructing a plastic model kit or instructions on how to treat a disease are fuzzy algorithms. Most readers with a solid background in computability theory will agree that these activities are not precisely defined; nevertheless, very few, if any, will think that they actually are algorithms. By employing a similar train of thought, Carol E. Cleland [32] has argued that cooking recipes are algorithms. In my own opinion, cooking recipes cannot be viewed as crisp algorithms, but maybe one can view them as some sort of fuzzy algorithms. Cleland has concluded that "Turing machines may be characterized as providing procedure schemas, i.e., temporally ordered frameworks for procedure." In addition, she has claimed that "Turing machine instructions cannot be said to prescribe actions, let alone *precisely* describe them." Furthermore, Cleland has argued against the idea that Turing machine "symbols" are genuine symbols. Thus, in a way, she had argued that Turing machines are implicitly vague, but obviously, I am interested in the direct introduction of vagueness, which is the subject of this chapter.

One simple way to turn a fuzzy command into a crisp one is to specify the fuzzy sets that are involved. The fuzzy command above uses two fuzzy sets. First it uses a fuzzy set that corresponds to the class of numbers that are approximately equal to 10, and, second, it uses a fuzzy set that corresponds to the class of numbers that are approximately equal to 5. Thus,

A. Syropoulos, *Theory of Fuzzy Computation*, IFSR International Series on Systems Science and Engineering 31, DOI 10.1007/978-1-4614-8379-3_4,
© Springer Science+Business Media New York 2014

the meaning of the fuzzy command above can be made precise by specifying the two fuzzy sets. Although there is a way to assign a precise meaning to a sequence of fuzzy commands, still it is not clear how these commands must be executed.

The most "obvious" way to execute fuzzy commands is to associate their fuzzy sets with collections of probabilities that are proportional to membership degrees and then to employ some probabilistic method to execute these commands. However, this approach is deeply flawed since it implicitly assumes that fuzziness and probability theory are almost the same thing. A second approach is to arbitrarily choose an α-cut of the fuzzy sets, which would make any value that belongs to the α-cut a permissible value. One can define ways to handle specials cases, but as Zadeh [145, p. 98] confesses, "the several particular cases considered above by no means provide definitive answers to the questions related to the execution of fuzzy instructions."

Zadeh had realized that his fuzzy algorithms were not rigorously defined. Thus, he proposed that one should define a fuzzy variant of the Turing machine in order to characterize fuzzy algorithms. He had noted that the state of a Turing machine at time $n + 1$ should depend on the state and the symbol being scanned at time n, that is,

$$q^{n+1} = f(q^n, S^n),$$

where q^n and S^n are variables ranging over the set of states and the set of tape symbols, respectively. This implies that (q^{n+1}, q^n, S^n) is an element of a subset of $Q \times Q \times \Gamma$. In the fuzzy setting, this relation should be replaced by a fuzzy one. Thus, each triple (q^{n+1}, q^n, S^n) is associated with a feasibility degree. In different words, there is a feasibility degree to which the machine will enter state q^{n+1} when in state q_n and the scanning device has read the symbol S^n.

In conclusion, every fuzzy Turing machine uniquely defines a fuzzy algorithm. However, Zadeh had realized that "the identification of an algorithm with a Turing machine restricts the applicability of the notion of an algorithm, whether fuzzy or not, to those situations in which the variables entering into the algorithm range over finite, or, at most countable sets."

4.2   Fuzzy Formal Languages and Grammars

Any Turing machine accepts a formal language. Similarly, any fuzzy Turing machine, an intuitive description of which has just given, should be able to accept a fuzzy language. But what is a fuzzy language and what does makes a language a fuzzy one? The concept of fuzzy languages was introduced by E.T. Lee and Zadeh [82] (see also [95] for a modern account of the field):

Definition 4.2.1 A fuzzy language λ over an alphabet S (i.e., an ordinary set of symbols) is a fuzzy subset of S^*.

If $s \in S^*$, then $\lambda(s)$ is the grade of membership that s is a member of the language.

Example 4.2.1 Consider the following set that includes all the sequences of zeros followed by ones:

$$L = \left\{ 0^i 1^j \mid i \neq j \text{ and } i, j > 0 \right\}.$$

Then the following function

$$\lambda(0^i 1^j) = \begin{cases} j/i, & \text{if } i > j, \\ i/j, & \text{otherwise} \end{cases}$$

defines a fuzzy, though peculiar, language.

Assume that λ_1 and λ_2 are two fuzzy languages over T. The *union* of λ_1 and λ_2 is the fuzzy language denoted by $\lambda_1 \cup \lambda_2$ and defined by

$$(\lambda_1 \cup \lambda_2)(x) = \lambda_1(x) \vee \lambda_2(x), \ \forall x \in T^*.$$

The *intersection* of λ_1 and λ_2 is the fuzzy language denoted by $\lambda_1 \cap \lambda_2$ and defined by

$$(\lambda_1 \cap \lambda_2)(x) = \lambda_1(x) \wedge \lambda_2(x), \ \forall x \in T^*.$$

The *concatenation* of λ_1 and λ_2 is the fuzzy language denoted by $\lambda_1 \lambda_2$ and defined by

$$(\lambda_1 \lambda_2)(x) = \bigvee_u \Big\{ \lambda_1(u) \wedge \lambda_2(v) \ \Big| \ x = uv, \ u, v \in T^* \Big\}.$$

By definition $\lambda^2(x) = (\lambda\lambda)(x)$, $\lambda^3(x) = (\lambda\lambda\lambda)(x)$, $\lambda^4(x) = (\lambda\lambda\lambda\lambda)(x)$, etc.

Suppose that λ is a fuzzy language in T. Then the fuzzy subset λ^* of T^* defined by

$$\lambda^* = \bigvee \Big\{ \lambda^n(x) \ \Big| \ n = 0, 1, 2, \dots \Big\}$$

for all $x \in T^*$ is called the *Kleene closure* of λ.

Roughly, a *fuzzy grammar* is a set of rules for generating the elements of a fuzzy language. In particular, a fuzzy grammar is a quadruple (V_N, V_T, P, S) where V_T and V_N are disjoint sets of terminal and nonterminal (syntactic class) symbols, respectively, P is a set of fuzzy production rules, and $S \in V_N$ is the starting symbol. The elements of P are expressions of the form

$$\alpha \xrightarrow{\rho} \beta, \rho > 0,$$

where $\alpha, \beta \in (V_T \cup V_N)^*$ and ρ is the plausibility degree that α can generate β.

A fuzzy grammar G generates a fuzzy language $L(G)$ in the following manner. A string ξ of terminal symbols belongs to $L(G)$ if and only if ξ is derivable from S. The plausibility degree to which ξ belongs to $L(G)$ is

$$L(G)(\xi) = \bigvee \min\{\rho_0, \rho_1, \dots, \rho_m\}, \tag{4.1}$$

where $S \xrightarrow{\rho_0} \alpha_1, \alpha_1 \xrightarrow{\rho_1} \alpha_2, \dots, \alpha_m \xrightarrow{\rho_m} \xi$. Note that the join is taken over all derivation chains from S to ξ, where a *derivation chain* from α_1 to α_m is an expression like the following one:

$$\alpha_1 \xrightarrow{\rho_1} \alpha_2 \cdots \alpha_{m-1} \xrightarrow{\rho_{m-1}} \alpha_m.$$

Fuzzy formal languages can form the basis of a theory of molecular computing [108]. In order to see what this means, let me give an example. Typically, in DNA computing [102] computation is carried out by letting DNA strands to evolve in chemical solutions. Problems are encoded as DNA strands and their solutions are DNA strands, which are interpreted accordingly. Each DNA strand can be thought to be a string that consists of symbols drawn from a very small alphabet. In particular, one can define two sets of complementary sets N^+ and N^- of nucleotides:

$$N^- = \{A, T, C, G, U\} \text{ and } N^+ = \{a, t, c, g, u\},$$

where A and a stand for adenine, T and t for thymine, C and c for cytosine, G and g for guanine, and U and u for uracil. Now, DNA strands are built by following the following (crisp) matching matrix:

	a	t	c	g	u
A	0	1	0	0	0
T	1	0	0	0	0
C	0	0	0	1	1
G	0	0	1	0	0
U	0	0	1	0	0

Typically, uracil replaces thymine in RNA macromolecules and is usually not found in DNA strands. However, it is quite possible that the building of DNA strands has something vague; thus, one can device a matching matrix that is fuzzy:

	a	t	c	g	u
A	$m_{1,1}$	$m_{1,2}$	$m_{1,3}$	$m_{1,4}$	$m_{1,5}$
T	$m_{2,1}$	$m_{2,2}$	$m_{2,3}$	$m_{2,4}$	$m_{2,5}$
C	$m_{3,1}$	$m_{3,2}$	$m_{3,3}$	$m_{3,4}$	$m_{3,5}$
G	$m_{4,1}$	$m_{4,2}$	$m_{4,3}$	$m_{4,4}$	$m_{4,5}$
U	$m_{5,1}$	$m_{4,2}$	$m_{5,3}$	$m_{5,4}$	$m_{5,5}$

where $m_{i,i} = 0$ and $0 \leq m_{i,j} \leq 1$. By fuzzifying the matching process, one has a basic computational ingredient that is fuzzy. Clearly, the development of a theory of fuzzy DNA computing is an open research area.

4.3   Fuzzy Turing Machines

Santos [113] was the first researcher who had given a formal definition of a fuzzy Turing machine. His initial version of a fuzzy Turing machine was called *Turing fuzzy algorithm*, but it could not handle a number of problems as shown by Shi-Kuo Chang [23], who proposed an alternative model of *fuzzy programs*. Later on, Santos [116] proposed an even more general version of a fuzzy Turing machine, which he called *Turing W-machine*, that encompasses the functionality of all previous attempts.

4.3.1   Turing Fuzzy "Algorithms"

Santos gave a definition of fuzzy Turing machine which is a fuzzy extension of the one provided by Vladeta Vučković [138]. Vučković called his machine a Turing algorithm and so quite naturally Santos called his machine a Turing fuzzy algorithm. Santos's definition follows:

Definition 4.3.1 A fuzzy Turing machine Z is a quintuple (A, B, Q, p, h) where

- A is the input alphabet;

- B is an *auxiliary* alphabet that may contain, among others, the symbol ⌣ and the symbol *, which is used to write n-tuples of words;

- Q is the set of internal states;

- $p : Q \times U \times V \times Q \to [0, 1]$ is the transition function, where $U = A \cup B$, $V = U \cup \{L, N, R\}$, $L, N, R \notin U$, and for every $u \in U$, $p(q, u, 0, q') = 0$ if $q \neq q'$; also $p(q, u, u', q')$ denotes the degree to which the machine may enter state q' and perform the action "specified" by u' when the present state is q and the symbol scanned is u; and

- $h : Q \to [0, 1]$ is the *initial distribution*.

Given a fuzzy Turing machine Z an expression α of Z is a string of symbols chosen from $U \cup Q$. Also, an expression α is called a *tape expression* when $\alpha \in U^*$. In addition, α is an *instantaneous expression* of Z if and only if it contains exactly one $q \in Q$ and q is not the rightmost symbol.

Definition 4.3.2 For every pair of instantaneous expressions α and β of Z, we define the plausibility that Z may leave the instantaneous expression α and enter β as follows:

$$q_Z(\alpha, \beta) = \begin{cases} p(q, u, u', q') & \text{if } \alpha = \sigma q u \tau, \beta = \sigma q' u' \tau, u' \in U, \\ p(q, u, R, q') & \text{if } \alpha = \sigma q u u' \tau, \beta = \sigma u q' u' \tau, u' \in U \\ & \text{or } \alpha = \sigma q u, \beta = \sigma u q' \, ⌣, \\ p(q, u, L, q') & \text{if } \alpha = \sigma u' q u \tau, \beta = \sigma q' u' u \tau, u' \in U \\ & \text{or } \alpha = q u \tau, \beta = q' \, ⌣ u \tau, \\ 0 & \text{otherwise,} \end{cases}$$

where σ and τ are (possibly empty) tape expressions of Z.

Not so surprisingly, Santos defined the function q_Z by "borrowing" the definition of the function that defines the probability that a probabilistic Turing machine may leave the instantaneous expression α and enter β (compare this definition with Definition 2.2.4 and the discussion that follows that definition). Function q_Z can be readily extended to $q_Z^{(n)}$, $n = 0, 1, 2, ...$, as follows:

$$q_Z^{(0)}(\alpha, \beta) = \begin{cases} 1, & \text{if } \alpha = \beta, \\ 0, & \text{if } \alpha \neq \beta, \end{cases}$$

$$q_Z^{(n+1)}(\alpha, \beta) = \bigvee_\gamma \left\{ \min\left\{ q_Z^{(n)}(\alpha, \gamma), q_Z(\gamma, \beta) \right\} \right\}.$$

In a sense, this machine behaves like the *maximin sequential-like* machines of Santos [111].

Definition 4.3.3 For every pair of instantaneous expressions α and β of a fuzzy Turing machine Z we define

$$t_Z^{(n)}(\alpha, \beta) = \min\{p(q, u, N, q), q_Z^{(n-1)}(\alpha, \beta)\},$$

where q is the state of Z at β and u is the symbol scanned by Z at β. In addition, we define

$$t_Z(\alpha, \beta) = \bigvee_n t_Z^{(n)}(\alpha, \beta).$$

$t_Z^{(n)}(\alpha, \beta)$ is interpreted as the degree to which "after n steps" an *algorithmic procedure* will "terminate" with β, given that it starts with α. $t_Z(\alpha, \beta)$ is interpreted as the degree to which the algorithmic procedure that starts with α will "terminate" with β after a finite but indefinite number of steps.

Each n-tuple $(w_1, w_2, ..., w_n)$ of words of a fuzzy Turing machine Z is associated to the tape expression $\overline{(w_1, w_2, ..., w_n)} = w_1 * w_2 * \cdots * w_n$, where each empty w_i is replaced by ⌴. Also, if α is an expression of Z, then $\langle \alpha \rangle$ is a word of Z obtained from α by removing all symbols not belonging to A.

Definition 4.3.4 An n-ary fuzzy word function in A is a function from the collection of all $(n + 1)$-tuples of words into $[0, 1]$.

Definition 4.3.5 For every fuzzy Turing machine Z and for each integer n, an n-ary fuzzy word function $\Phi_Z^{(i)}$ in A (i.e., the input alphabet) is associated as follows:

$$\Phi_Z^{(n)}(w_1, w_2, ..., w_n, w) = \bigvee_{\langle \beta \rangle = w} \bigvee_{q \in Q} \{\min\{h(q), t_Z(\alpha, \beta)\}\},$$

where $\alpha = \overline{q(w_1, w_2, ..., w_n)}$ and the first least upper bound is taken over all instantaneous expressions β such that $\langle \beta \rangle = w$.

The value of $\Phi_Z^{(n)}(w_1, w_2, ..., w_n, w)$ represents the degree to which the fuzzy Turing machine Z that starts with tape $\overline{(w_1, w_2, ..., w_n)}$ will terminate with tape w. The following follows immediately from these definitions:

Theorem 4.3.1 *For every fuzzy Turing machine $Z = (A, B, Q, p, h)$, there exists a fuzzy Turing machine $Z_0 = (A, B, Q_0, p_0, h_0)$, where $Q_0 = S \cup \{q_0\}$, $q_0 \notin A \cup Q \cup B$, and*

$$p_0(q', u, v, q'') = \begin{cases} p(q', u, v, q''), & \text{if } q', q'' \in Q, \\ h(q''), & \text{if } q' = q_0 \text{ and } v = u \in U, \\ 0, & \text{otherwise,} \end{cases}$$

$$h_0(q) = \begin{cases} 1, & \text{if } q = q_0, \\ 0, & \text{if } q \neq q_0 \end{cases}$$

such that

$$\Phi_Z^{(n)}(w_1, w_2, ..., w_n, w) = \Phi_{Z_0}^{(n)}(w_1, w_2, ..., w_n, w),$$

for all n and $w_1, w_2, ..., w_n, w \in A^*$.

In different words, there is no loss of generality in considering a fuzzy Turing machine for which $h(q_0) = 1$ and $h(q) = 0$ whenever $q \neq q_0$. In cases like this, q_0 is the initial state.

4.3.2   Fuzzy Programs

Chang's Approach

Chang's [23] model of fuzzy computation consists of three parts: a machine that performs the computations, a fuzzy program that controls the machine, and the relationship between the machine and the fuzzy program. The machine is a finite-state machine and the fuzzy program is a *regular expression* (i.e., the kind of expressions used to perform string searches and modifications in programming languages like Perl). The relationship between the machine and the fuzzy program is determined by two functions—the *feasibility* function f and the *performance* function λ.

Definition 4.3.6 A *finite-state* machine $\mathcal{M}$ is a quintuple $(K, Q, \Psi, q_{\text{init}}, T)$ where

- K is a finite nonempty set that contains the machine instructions;

- Q is finite nonempty set of the internal states;

- $q_{\text{init}} \in Q$ is the initial state;

- $T \subset Q$ is the set of final states; and

- $\Psi : Q \times K \to Q$ is the transition function.

It is easy to extend function Ψ so it can handle strings of instructions:

$$\Psi(q, \varepsilon) = q$$
$$\Psi(q, i_1 i_2 \cdots i_{n-1} i_n) = \Psi(\Psi(q, i_1 i_2 \cdots i_{n-1}), i_n),$$

where $q \in Q$, $i_k \in K$, and ε denotes the empty string. A string of instructions $i_1 i_2 \cdots i_{n-1} i_n$ is a *machine program* if

$$\Psi(q_{\text{init}}, i_1 i_2 \cdots i_{n-1} i_n) = \Psi(\cdots \Psi(\Psi(q_{\text{init}}, i_1), i_2), \cdots, i_n) \in T.$$

Definition 4.3.7 A fuzzy machine $\mathcal{N}$ is a quadruple $(\Sigma, \mathcal{M}, f, \lambda)$ where

- Σ is the finite and nonempty set of fuzzy instructions;

- $\mathcal{M}$ is a finite-state machine;

- $f : Q \times \Sigma \times K \to \{0, 1\}$ is the feasibility function; and

- $\lambda : Q \times \Sigma \times K \to [0, 1]$ is the performance function.

The value of $f(q, \sigma, i)$ specifies whether the instruction i can be executed when the machine is in state q and the fuzzy instruction σ has been received. In particular, if $f(q, \sigma, i) = 1$, then the instruction i can be executed; otherwise it cannot be executed. Also, the value of the performance function gives a subjective performance evaluation that is used by the machine to choose which machine instruction to execute in a particular case. For example, if $\lambda(q, \sigma, i_1) > \lambda(q, \sigma, i_2)$ and $\mathcal{M}$ is in state q and it has received the fuzzy instruction σ, then it is more likely that the next instruction to be executed will be i_1.

In the most general case, a fuzzy program is a regular expression over Σ,[1] while an *elementary* fuzzy program is a string in Σ^*. For example, consider the fuzzy command which we encountered in p. 51: *make y approximately equal to* 10. The intended meaning of this command is *make y equal to* 9.8 *or* 9.9 *or...., or* 10.1*, or....* Now, assume that σ means *make y equal to* 9.8, $\sigma\sigma$ means *make y equal to* 9.81, $\sigma\sigma\sigma$ means *make y equal to* 9.82, etc. Then the regular expression $\sigma*$ would convey the required meaning. Obviously, this schema works better for fuzzy commands like *increase y by several (integral) units*, thus revealing a weakness of this approach. Moreover, the fuzzy instruction "execute α or β" is represented by $\alpha \mid \beta$.

Definition 4.3.8 Assume that $m = i_1 i_2 \cdots i_n$ is a machine program of $\mathcal{M}$ and that $q_{\text{init}}, q_1,$ ..., q_n are the successive states of $\mathcal{M}$ when this machine executes m (i.e., $q_j = \Psi(q_{j-1}, \sigma_j, i_j)$) for $1 \le j \le n$). Then, the machine program m is said to be an *execution* of the elementary fuzzy program $\varphi = \sigma_1 \sigma_2, ..., \sigma_n$ if $f(q_{j-1}, \sigma_j, i_j) = 1, 1 \le j \le n$. In addition, a fuzzy program ρ is *executable* with respect to the fuzzy machine $\mathcal{N}$ if there is an elementary fuzzy program φ in ρ and a machine program m such that m is an execution of φ.

Now let us see what it means to execute a elementary fuzzy program. The first question is how to choose from among several feasible instructions. Obviously, a fuzzy machine $\mathcal{N}$ should choose the instructions in a way that ensures the best overall performance.

Definition 4.3.9 Assume that $m = i_1 i_2 \cdots i_n$ is an execution of an elementary fuzzy program $\varphi = \sigma_1 \sigma_2, ..., \sigma_n$ in a fuzzy program ρ with respect to a fuzzy machine $\mathcal{N}$. Also, let $q_{\text{init}}, q_1, ..., q_n$ be the successive states of $\mathcal{M}$ when it executes m. In addition, assume that $\hat{\lambda} : (\Sigma \times K)^* \to I$ is a function defined as follows:

$$\hat{\lambda}(\varphi, m) = \left\{ \begin{array}{ll} \min_{1 \le j \le n}\{\lambda(q_{j-1}, \sigma_j, i_j)\}, & \text{if } m \text{ is an execution of } \varphi, \\ 0, & \text{otherwise.} \end{array} \right.$$

Then, the *most preferable execution* of an elementary fuzzy program φ, denoted by m_{mp}, is the machine program in D_φ such that

$$\hat{\lambda}(\varphi, m_{\text{mp}}) \ge \hat{\lambda}(\varphi, m)$$

for all the other machine programs m in D_φ, where D_φ denotes the set of machine programs of $\mathcal{N}$ that are execution of an elementary fuzzy program in a fuzzy program ρ. Similarly, the most preferable execution m_{mp} if a fuzzy program ρ is the machine program in D_ρ such that there is an elementary fuzzy program φ_{mp} in E_ρ and

$$\hat{\lambda}(\varphi_{\text{mp}}, m_{\text{mp}}) \ge \hat{\lambda}(\varphi, m)$$

for any φ in E_ρ and any m in D_ρ and where E_ρ denotes the set of executable elementary fuzzy programs of regular fuzzy program ρ.

---

1. Donald M. Kaplan [71] had used regular expressions to represent elementary programs in order to study program equivalence.

The following immediate result is quite important:

Corollary 4.3.1 *The most preferable execution of a regular fuzzy program can be found effectively (i.e., one can construct an algorithm to find it).*

Now let us say how to execute elementary fuzzy programs:

Definition 4.3.10 The *simple execution* of an elementary fuzzy program $\varphi = \sigma_1 \sigma_2, ..., \sigma_n$ consists of the following steps:

(i) $j := 1$ and $q := q_{\text{init}}$.

(ii) Select the machine instruction $i \in K$ such that

$$f(q, \sigma_j, i) = 1 \text{ and } \lambda(q, \sigma_j, i) \geq \lambda(q, \sigma_j, i') \times f(q, \sigma_j, i')$$

for all other $i' \in K$. If the selection is possible, then goto (iii). Otherwise goto (vi).

(iii) $i_j := i$ and $q := \Psi(q, i)$.

(iv) If $(j = n) \wedge (q \in T)$, goto (v). If $(j = n) \wedge (q \notin T)$, goto (vi). Otherwise $j := j + 1$ and goto (ii).

(v) Success exit. The execution is $i_1 i_2 \cdots i_n$.

(vi) Failure exit.

When the performance function of a fuzzy machine is not known (e.g., because more than one definitions is available), one needs a mechanism to construct a performance function λ such that an execution of a given fuzzy program can be found with minimum cost. The following *modified* simple execution of a fuzzy program does exactly what is needed:

Definition 4.3.11 Given an elementary fuzzy program $\varphi = \sigma_1 \sigma_2, ..., \sigma_n$, the *modified* simple execution of it consists of the following steps:

(i) $j := 1$ and $q_0 := q_{\text{init}}$.

(ii) $K_j := K$, where K_j is a set.

(iii) Select the machine instruction $i \in K_j$ such that

$$f(q_{j-1}, \sigma_j, i) = 1 \text{ and } \lambda(q_{j-1}, \sigma_j, i) \geq \lambda(q_{j-1}, \sigma_j, i') \times f(q_{j-1}, \sigma_j, i')$$

for all other $i' \in K_j$. If the selection is possible, continue. Otherwise goto (vii).

(iv) $i_j := i$, $q_j := \Psi(q_{j-1}, i)$, and $K_j := K_j \setminus \{i\}$.

(v) If $(j = n) \wedge (q_j \in T)$, goto (vi). If $(j = n) \wedge (q_j \notin T)$, goto (iii). Otherwise $j := j + 1$ and goto (ii).

(vi) Success exit. The execution is $i_1 i_2 \cdots i_n$.

(vii) $j := j - 1$. If $j = 0$, failure exit. Otherwise goto (iii).

Tanaka–Masaharu's Approach

A extension of Chang's fuzzy machine has been presented by Kokichi Tanaka and Masaharu Mizumoto [131]. This extended fuzzy machine is based on a generalization of Chang's machine:

Definition 4.3.12 A *generalized machine* $\mathcal{M}$ is a sextuple $(K, Q, \Psi, q_{\text{init}}, T, V)$ where

- K is a finite set of machine instructions;

- Q is a finite set of internal states;

- $\Psi : Q \times K \times Q \to V$ is the *state transition function* whose return value designates the plausibility degree for which when a particular machine instruction i is selected for execution the machine can go from state q to state q';

- $q_{\text{init}} \in Q$ is an initial state;

- $T \subset Q$ is a finite and nonempty set of final states; and

- V is a "space" of grades that control the state transition.

Typically, the "space" V is a *lattice ordered semigroup*, that is, a quintuple $(A, \vee, *, 0, 1)$, where A is a lattice (i.e., a partially ordered set in which each two-element subset has both a least upper and a greatest lower bound) for which the following identities hold

$$c * (a \vee b) = c * a \vee c * b \text{ and } (a \vee b) * c = a * c \vee b * c.$$

Also, 0 and 1 are the zero (least element) and the identity (greatest element). Now, the state transition function can be defined as follows. Assume that $m = i_1 i_2 \cdots i_n$ and $m \in K^*$ (i.e., m is a finite sequence of machine instructions). Also assume that at each execution step the state transition function will be $\Psi(q_{\text{init}}, i_1, q_1), \Psi(q_1, i_2, q_2), \ldots, \Psi(q_{n-1}, i_n, q_n)$. Then, the state of the machine changes from q_{init} to q_n in a number of steps that are dictated by the machine instructions m and the degree that corresponds to this change is given by

$$\Psi(q_{\text{init}}, i_1, q_1) * \Psi(q_1, i_2, q_2) * \cdots * \Psi(q_{n-1}, i_n, q_n).$$

Using this result, one can easily extend the definition of the state transition function to handle sequences of machine instructions as follows:

$$\Psi(q, \varepsilon, q') = \begin{cases} 1, & \text{if } q = q', \\ 0, & \text{if } q \neq q', \end{cases}$$

$$\Psi(q, m, q') = \bigvee_{q_{\text{init}}, q_2, \ldots, q_n} \left\{ \Psi(q_{\text{init}}, i_1, q_1) * \Psi(q_1, i_2, q_2) * \cdots * \Psi(q_{n-1}, i_n, q_n) \right\}.$$

The definition of the generalized fuzzy machines that are associated with the generalized machines just defined is given below:

Definition 4.3.13 A *generalized fuzzy machine* $\mathcal{N}$ is a triple $(\Sigma, \mathcal{M}, W)$ where

- each fuzzy instruction σ_j is a function $\sigma_j : Q \times K \to W$ and Σ is a finite set of fuzzy instructions;

- $\mathcal{M}$ is a generalized machine; and

- the elements of space W are plausibility degrees that express the plausibility of selecting the machine instruction i_j when $\mathcal{M}$ is in state q_j and has received the fuzzy instruction σ_j.

The following definition describes how the generalized fuzzy machine operates.

Definition 4.3.14 Given an elementary fuzzy program $\varphi = \sigma_1\sigma_2, ..., \sigma_n$, the execution procedure of φ by the generalized fuzzy machine consists of the following steps:

(i) $j := 1$ and $q_0 := q_{\text{init}}$.

(ii) $K' := \left\{ i \mid \sigma_j(q_{j-1}, i) > \theta \right\}$ and $K_{j, q_{j-1}} := K'$.

(iii) If $K_{j, q_{j-1}} \neq \varnothing$, goto (iv). Otherwise, set $i := i - 1$. If $i = 0$, failure exit.

(iv) Select the machine instruction $i \in K_{j, q_{j-1}}$ such that

$$\sigma_j(q_{j-1}, i) \geq \sigma_j(q_{j-1}, i')$$

for all other $i' \in K$.

(v) $i_j := i$, $K_{j, q_{j-1}} := K_{j, q_{j-1}} \setminus \{i\}$, $Q' := \left\{ q \mid \Psi(q_{j-1}, i_j, q) > \theta \right\}$, and $Q_{j, q_{j-1}, i_j} := Q'$.

(vi) If $Q_{j, q_{j-1}, i_j} = \varnothing$, goto (iii). Otherwise, continue.

(vii) Select the state of the machine transits from q_{i-1} to q such that

$$\Psi(q_{i-1}, i_j, q) \geq \Psi(q_{i-1}, i_j, q')$$

for all other $q' \in Q_{j, q_{j-1}, i_j}$.

(viii) $q_j := q$ and $Q_{j, q_{j-1}, i_j} := Q_{j, q_{j-1}, i_j} \setminus \{q\}$.

(ix) If $j \neq n$, set $j := j + 1$ and goto (ii). Otherwise, continue.

(x) If $q_j \notin T$, goto (vi). Otherwise, success exit. The execution is $i_1 i_2 \cdots i_n$.

Santos's Approach

The work of Tanaka and Mizumoto had revealed that it is possible to broaden the notion of a fuzzy program. Indeed, Santos [116] had proposed a formulation of fuzzy programs that was general enough to encompass all formulations known to him. At the same time, this new formulation could be used to deal with problems of practical interest.

Santos's formulation of a fuzzy program is based on earlier work he did on the notion of a program [115]. In turn, this work was based on work done by Dana Scott [117]. Scott, and consequently Santos, had defined a program to be a sequence of commands:[2]

Definition 4.3.15 A command is a string of one of the following forms:

$$\text{Start: goto } L$$
$$L: \text{do } F; \text{goto } L_1$$
$$L: \text{if } P \text{ then goto } (L_1, L_2, ..., L_n)$$
$$L: \text{halt}$$

where $L_i \in \mathscr{L}$ (a set of labels), $F \in \mathfrak{F}$ (the set of function or operation symbols), and $P \in \mathfrak{P}_n$ (the set of n-valued predicate or test symbols). These commands are called start, operation, test, and halt commands. Also, a program contains exactly one start command while all command labels must be different.

The "goto L" command just transfers control to the command with label L. Also, the "halt" commands aborts program execution. The sets $\mathfrak{F}$ and $\mathfrak{P}_n$ have not been defined, but for the moment it suffices to say that one can use these sets of symbols together with some other sets to define machines. For example, as was demonstrated by Scott [117], the Turing machine can be defined this way. In this case, the set $\mathfrak{F}$ consists of the operations MOVERIGHT, MOVELEFT, PRINT, and ERASE, while one of the elements of $\mathfrak{P}_n$ is the predicate IS BLANK?. The "if" command looks like the arithmetic IF command of FORTRAN, only in this case P returns a positive integer and depending on the return value it transfers control to the command whose label is $L_1, L_2, ..., $ or L_n.

In order to define his models of fuzzy programs, Santos used the notion of an *ordered semirings*. As was suggested to me by some reviewer, if the theory of t-norms and t-conorms was available to Santos, he would have preferred to use t-norms and t-conorms in his work instead of ordered semirings.

Definition 4.3.16 An ordered semiring is a quadruple $(W, \oplus, \otimes, \prec)$, where W is an nonempty set, $\oplus$ and $\otimes$ are binary operations on W, and $\prec$ is a partial ordering relation on W satisfying the following conditions:

(i) for all $a, b, c \in W$

$$a \oplus b = b \oplus a$$
$$a \otimes b = b \otimes a$$
$$a \oplus (b \oplus c) = (a \oplus b) \oplus c$$
$$a \otimes (b \otimes c) = (a \otimes b) \otimes c$$

and

$$a \otimes (b \oplus c) = (a \otimes b) \oplus (a \otimes c);$$

---

2. In fact, Santos, following Scott, had defined a program to be a set of commands. Note that typically an ordinary program is understood to be a sequence of commands to be executed in that particular order. But since there is no implicit order in the way one may pick up elements from any set, it is better to talk about sequences than sets.

(ii) there exist $0, 1 \in W$ such that $a \oplus 0 = 0 \oplus a = a$ and $a \otimes 1 = 1 \otimes a = a$ for all $a \in W$; and

(iii) for every $a, b \in W$, $a \prec a \oplus b$ and $b \prec a \oplus b$.

According to Santos, the following ordered semirings yield interesting models of fuzzy programs: $W_W = (\mathbb{R}^+, +, \times, \leq)$, where $\mathbb{R}^+$ is the set of positive real numbers including zero, $W_X = ([0, 1], \max, \min, \leq)$, $W_I = ([0, 1], \min, \max, \geq)$, $W_T = ([0, 1], \max, \times, \leq)$, and $W_N = (\{0, 1\}, \max, \min, \leq)$. For reasons of brevity, from now on I will use the symbol W to denote an ordered semiring $(W, \oplus, \otimes, \prec)$.

Definition 4.3.17 Assume that W is an ordered semiring and that U and V are nonempty sets. Then, a W-function f from U to V is a function from $V \times U$ to W.

Note that the degree to which the value of a function f at u is v is $f(v \mid u)$.

Definition 4.3.18 Assume that W is an ordered semiring and that the symbols I and O, stand for *input* and *output*, respectively, do not belong to either $\mathfrak{F}$ or $\mathfrak{P}$. A W-machine is a function $\mathfrak{M}$ defined on the set $\{I\} \cup \mathfrak{F} \cup \mathfrak{P} \cup \{O\}$ such that the sets X (input set), M (memory set), and Y (output set) exist and

(i) $\mathfrak{M}_I$ is a W-function from X to M;

(ii) for all $F \in \mathfrak{F}$, $\mathfrak{M}_F$ is a W-function from M to M;

(iii) for all $P \in \mathfrak{P}$, there is a positive integer n such that $\mathfrak{M}_P$ is a W-function from M to $\{1, 2, ..., n\}$; and

(iv) $\mathfrak{M}_O$ is a W-function from M to Y.

These four kinds of functions are called input, operation, test, and output functions, respectively.

When W is W_W, W_X, W_I, W_T, or W_N, then the corresponding W-machines are called weighted, maximin, minimax, max-product, or nondeterministic machines, respectively.

Definition 4.3.19 Assume that $\mathfrak{M}$ is a W-machine and n a positive integer. Then, $\mathfrak{P}_n$ is the subset of $\mathfrak{P}$ consisting of all $P \in \mathfrak{P}$ such that $\mathfrak{M}_P$ is a W-function from M into $\{1, 2, ..., n\}$.

Definition 4.3.20 Assume that π is a program and $\mathfrak{M}$ a W-machine. Program π is *admissible* on $\mathfrak{M}$, or $\mathfrak{M}$ admits π, iff

(i) for all operation commands "L: do F; goto L'," $F \in \mathfrak{F}$; and

(ii) for all test commands "L: if P then goto $(L_1, L_2, ..., L_n)$," $P \in \mathfrak{P}_n$.

Assume that π is a program and $\mathfrak{M}$ a W-machine that admits π. For all $L_1, L_2 \in \mathfrak{L}$, $m_1, m_2 \in M$, and $w \in W$, the expression $(L_1, m_1) \xrightarrow{w} (L_2, m_2)$ denotes that either a command of the form

$$L_1 \colon \text{do } F; \text{goto } L_3$$

belongs to π, which implies that $L_2 = L_3$ and $w = \mathfrak{M}_F(m_2 \mid m_1)$, or a command of the form

$$L_1 \colon \text{if } P \text{ then goto } (L_1', L_2', ..., L_n')$$

belongs to π, which implies that $m_1 = m_2, L_2 = L_i$ for some $i = 1, 2, ..., n$, and $w = \mathfrak{M}_P(i \mid m)$. Otherwise, $(L_1, m_1) \xrightarrow{0} (L_2, m_2)$ where 0 is the identity of the $\oplus$ operation. Now let us see what computation is:

Definition 4.3.21 Assume that $\mathfrak{M}$ admits π. Then a computation by π on $\mathfrak{M}$ is a finite sequence

$$x, L_0, m_0, L_1, m_1, ..., L_n, m_n, y \tag{i}$$

where $x \in X, y \in Y, L_i \in \mathscr{L}, m_i \in M, i = 0, 1, ..., n$; and L_0 is the label where control is transferred in the start command of π, and L_n is the label of some halt command in π. The plausibility degree associated with (i) is an element $w \in W$ such that $w = w_0 \otimes w_1 \otimes w_2 \otimes \cdots \otimes w_{n+1}$, where $w_i \in W, i = 1, 2, ..., n, w_0 = \mathfrak{M}_I(m_0 \mid x), w_{n+1} = \mathfrak{M}_O(y \mid m_n)$, and for every $i = 1, 2, ..., n, (L_{i-1}, m_{i-1}) \xrightarrow{w_i} (L_i, m_i)$. The computation (i) is feasible if and only if its plausibility degree $w \neq 0$. The x and y in (i) are called the input and the output of the computation, respectively.

Now that we have defined the notion of a fuzzy program à la Santos, we need to define a machine capable to execute such programs. He started with a modified version of the Tanaka–Mizumoto generalized machine described in Definition 4.3.12 on p. 60:

Definition 4.3.22 A *generalized machine* $\mathscr{M}$ is a sextuple $(K, Q, \Psi, q_{\text{init}}, T, V)$, where V is an ordered semiring and Ψ is a W-function from $K \times Q$ into Q.

Definition 4.3.23 Assume that $\mathscr{M}$ is a generalized machine. Then, the W-function Ψ can be extended to a $\mathscr{M}$-function from $K^* \times Q$ into Q as follows:

$$\Psi(q'' \mid \varepsilon, q') = \begin{cases} 1, & \text{if } q' = q'', \\ 0, & \text{if } q' \neq q'' \end{cases}$$

and

$$\Psi(q'' \mid ii_0, q') = \sum \left\{ \Psi(q'' \mid i, q) \otimes \Psi(q' \mid i_0, q) \,\middle|\, q \in Q \right\},$$

where $q', q'' \in Q, i \in K^*, i_0 \in K$, and

$$\sum \left\{ a_1, a_2, ..., a_n \right\} = a_1 \oplus a_2 \oplus \cdots \oplus a_n$$

for all $a_i \in V$.[3]

---

3. This last clarification is actually a proposition; nevertheless, its proof is trivial and not necessary for what comes next.

Santos's definition of a generalized fuzzy machine is similar but not identical to the one given by Tanaka and Mizumoto:

Definition 4.3.24 A generalized fuzzy machine is an ordered pair $(\mathcal{M}, \Sigma)$ where $\mathcal{M}$ is a generalized machine and Σ is a finite set of fuzzy instructions and each fuzzy command $\sigma \in \Sigma$ is a W-function from Q to K.

In Santos's approach, any element w of Σ^* is called an elementary fuzzy program and any function from Σ^* into V is called a fuzzy program. Here V is the ordered semiring of some generalized machine. Note that this definition of fuzzy programs is more general than the one given so far.

Definition 4.3.25 Assume that $(\mathcal{M}, \Sigma)$ is a generalized fuzzy machine, where $\mathcal{M}$ is the generalized machine $(K, Q, \Psi, q_{\text{init}}, T, V)$. Also, assume that $\Sigma^* \ni w = \sigma_1 \sigma_2 \cdots \sigma_n$. Then, an execution of w on $\mathcal{M}$ is a sequence

$$q_{\text{init}}, i_1, q_1, i_2, ..., i_n, q_n, \qquad (ii)$$

where $q_j \in Q$ and $i_j \in K$ for all $j = 1, 2, ..., n$ and $q_n \in T$. The plausibility degree associated with the execution of the sequence (ii) is an element of $v \in V$ such that $v = v_1 \otimes v_1' \otimes v_2 \otimes v_2' \otimes \cdots \otimes v_n \otimes v_n'$ where for every $j = 1, 2, ..., n$, $v_j = \sigma_j(i_j \mid q_{j-1})$ and $v_j' = \Psi(q_j \mid i_j, q_{j-1})$. The execution (ii) is feasible if and only if $v \neq 0$. The q_n in expression (ii) is called the final state of the execution.

Definition 4.3.26 Suppose that $(\mathcal{M}, \Sigma)$ is a generalized fuzzy machine. Then for every $w \in \Sigma^*$ and $q \in Q$, we define $A(q \mid w)$ to be $\sum V_0$, where V_0 is the set of plausibility degrees associated with some execution of w on A with final state q, if $q \in T$, and define $A(q \mid w) = 0$ if $q \notin T$.

Theorem 4.3.2 *There exists a W-machine $\mathfrak{M}$ where $X = Y = U^*$ for some set U, such that for every general fuzzy machine $(\mathcal{M}, \Sigma)$, where $\mathcal{M}$ is the generalized machine $(K, Q, \Psi, q_{\text{init}}, T, V)$, $\Sigma \subseteq U$, and $Q \subseteq U$, there exists a program π that is admissible on $\mathfrak{M}$ for which $\mathfrak{M}_\pi(q \mid w) = A(q \mid w)$ for all $q \in Q$ and $w \in \Sigma^*$.*

Proof Assume that $\mathfrak{M}$ is a W-machine with $X = Y = U^*$, where U is an arbitrary set, $M = U^* \times U^*$, $\mathscr{F} = \{F_0\} \cup \{F_u \mid c\}$, $\mathscr{P} = \{P_u \mid u \in U\} \cup \mathscr{P}_0$, and for every $\alpha, \beta, \gamma, \delta \in U^*$, and $t = 1, 2$

$$\mathfrak{M}_I\big((\alpha, \beta) \mid \gamma\big) = \begin{cases} 1, & \text{if } \alpha = \varepsilon \text{ and } \beta = \gamma, \\ 0, & \text{otherwise,} \end{cases}$$

$$\mathfrak{M}_O\big(\alpha \mid (\beta, \gamma)\big) = \begin{cases} 1, & \text{if } \alpha = \beta, \\ 0, & \text{otherwise,} \end{cases}$$

$$\mathfrak{M}_{F_0}\big((\alpha, \beta) \mid (\gamma, \delta)\big) = \begin{cases} 1, & \text{if } \alpha = \gamma \text{ and } \delta = u\beta \text{ for some } u \in U, \\ 0, & \text{otherwise,} \end{cases}$$

$$\mathfrak{M}_{F_u}\big((\alpha, \beta) \mid (\gamma, \delta)\big) = \begin{cases} 1, & \text{if } \beta = \delta \text{ and } \alpha = \gamma u, \\ 0, & \text{otherwise,} \end{cases}$$

$$\mathfrak{M}_{P_0}\big(t \mid (\alpha, \beta)\big) = \begin{cases} 1, & \text{if } t = 1 \text{ and } \beta = u\tau \text{ for some } \tau \in U^*, \\ & \text{or } t = 2 \text{ and } \beta \neq u\tau \text{ for all } \tau \in U^*, \\ 0, & \text{otherwise} \end{cases}$$

and $\mathscr{P}_0$ is the collection of all unconditional test functions. Except for the unconditional test functions, Santos [115] has defined a one-way finite automatoma in an almost identical way. Thus, $\mathfrak{M}$ can be considered as a one-way finite automaton. Assume that $\Sigma = \{\sigma_1, \sigma_2, ..., \sigma_n\}$, $K = \{i_1, i_2, ..., i_r\}$, $Q = \{q_{\text{init}}, q_1, ..., q_l\}$, and π is the following program:

$$
\begin{array}{rl}
\text{Start:} & \text{goto } L_0^1 \\
L_j^k: & \text{if } P_k \text{ then goto } (L_j^{k0}, L_j^{k+1}) \\
L_j^{k0}: & \text{do } F_0; \text{ goto } L_j^{k1} \\
L_j^{k1}: & \text{if } P_j^k \text{ then goto } (L_{1j}, L_{2j}, ..., L_{rj}) \\
L_{tj}: & \text{if } P_{tj} \text{ then goto } (L_0^1, L_1^1, ..., L_m^1) \\
L_\ell^{n+1}: & \text{do } F_{s_t}; \text{ goto } L \\
L: & \text{halt}
\end{array}
$$

where $k = 1, 2, ..., n$, $j = 0, 1, 2, ..., t$, $t = 1, 2, ..., r$, $\ell \in \{\ell \mid s_\ell \in T\}$, $\mathfrak{M}_{P_j^k}$ is a W-function from M to $\{1, 2, ..., r\}$ such that $\mathfrak{M}_{P_j^k}(z \mid m) = \sigma_i(i_z \mid q_j)$ for all $m \in M$ and $z = 1, 2, ..., r$, and $\mathfrak{M}_{P_{lj}}$ is a W-function from M to $\{1, 2, ..., n + 1\}$. Note that the instruction with label L_j^{k1} and L_{lj} simulates the action of σ_j and Ψ, respectively. Therefore, it is easy to verify that $\mathfrak{M}_\pi(q \mid w) = A(q \mid w)$ for all $w \in \Sigma^*$, and $q \in Q$, □

This result is a proof of the statement made above that the formulations of Chang and Tanaka–Mizumoto are special cases of the current formulation.

Other Approaches

Juan Luis Castro, Miguel Delgado, and Carlos Javier Mantas [21] presented an alternative formulation of fuzzy grammars. Unfortunately, these grammars are in fact fuzzy context sensitive grammars and so there is no new insight. Furthermore, the machines that are supposed to accept such languages are in a way special.

4.3.3   Turing W-Machines

Santos [116] has given another definition of a fuzzy Turing machine:

Definition 4.3.27 A Turing W-machine M is a septuple (U, V, S, Q, T, q_0, p) where

- U is the set of input symbols;

- V is the set of output symbols;

- S is the set of tape symbols;

- Q is the set of states;

- $T \subseteq Q$ is the set of terminal states;

- $q_0 \in Q$ is the initial state; and

- p is W-function from $T \times Q$ to $Q \times (S \cup \{R, L, N\})$, where $R, L, N \notin S$, such that for every $s \in S$, $p(q', N \mid s, q) = 0$ for $q \neq q'$.

In general, $p(q', z \mid c, q)$ expresses the plausibility of what happens next provided that the machine is in state q and the tape symbol c is scanned. In particular,

(i) if $z \in S^*$, then replace c by z and enter state q';

(ii) if $z = R$, then move one square to the right and enter state q';

(iii) if $z = L$, then move one square to the left and enter state q'; and

(iv) if $z = N$, then stop.

For the rest of the present discussion, it is assumed that $\sqcup \in S$ and $\sqcup \notin U \cup V$.

Definition 4.3.28 Assume that Z is a Turing W-machine and $\alpha \in (S \cup Q)^*$. Then, α is an instantaneous description of Z if and only if

(i) it contains exactly one $q \in Q$ and this symbol is not the rightmost symbol in α;

(ii) its leftmost symbol is not $\sqcup$; and

(iii) its rightmost symbol is not $\sqcup$ unless it is the symbol immediately to the right of q.

The collection of all instantaneous descriptions of Z will be denoted by $D(Z)$.

Definition 4.3.29 Assume that Z is Turing W-machine. Then p^Z is a W-function from $D(Z)$ into $D(Z)$ such that for every $\alpha, \beta \in D(Z)$:

$$p^Z(\beta\alpha) = \begin{cases} p(q',s' \mid s,q) & \text{if } \alpha = \zeta qs\delta, \quad \beta = \zeta q's'\delta, \quad s'\delta \neq \varepsilon \\ & \text{or } \alpha = \zeta qs, \quad \beta = \zeta q'\sqcup, \quad s' \neq \varepsilon, \\ p(q',R \mid s,q) & \text{if } \alpha = \zeta qss'\delta, \quad \beta = \zeta sq's'\delta, \quad \zeta s \neq \sqcup \\ & \text{or } \alpha = qss'\delta, \quad \beta = q's'\delta, \quad s = \sqcup \\ & \text{or } \alpha = \zeta qs, \quad \beta = \zeta sq'\sqcup, \quad \zeta s \neq \sqcup \\ & \text{or } \alpha = qs, \quad \beta = q'\sqcup, \quad s = \sqcup, \\ p(q',L \mid s,q) & \text{if } \alpha = \zeta s'qs\delta, \quad \beta = \zeta q's's\delta, \quad s\delta \neq \sqcup \\ & \text{or } \alpha = \zeta s'q\delta, \quad \beta = \zeta q's', \quad s = \sqcup \\ & \text{or } \alpha = qs\delta, \quad \beta = q'\sqcup s\delta, \quad s\delta \neq \sqcup \\ & \text{or } \alpha = qs, \quad \beta = q'\sqcup, \quad s = \sqcup, \\ 0 & \text{otherwise,} \end{cases}$$

where $\zeta, \delta \in S^*$, $q, q' \in Q$, and $s, s' \in S$.

Note that $p^Z(b \mid a)$ is the plausibility degree to which the instantaneous description a is followed by the instantaneous description b when the machine Z is running.

Definition 4.3.30 Assume that Z is a Turing W-machine. Then, a computation of Z with input $x \in U^*$ and output $y \in V^*$ is a finite sequence

$$\alpha_0, \alpha_1, \alpha_2, ..., \alpha_n, \tag{4.2}$$

of elements of $D(Z)$, where $\alpha_0 = q_0 x$, $\alpha_n = \beta q \gamma$ and where $\beta \gamma = y$, $q \in Q$. The plausibility degree associated with the computation (4.2) is

$$\omega = p^Z(\alpha_1 \mid \alpha_0) \otimes p^Z(\alpha_2 \mid \alpha_1) \otimes \cdots \otimes p^Z(\alpha_n \mid \alpha_{n-1}) \otimes p(q,N \mid s,q),$$

where s is the symbol contained in α_0 which is immediately to the right of q.

Definition 4.3.31 Assume that Z is a Turing W-machine. Then, for every $x \in U^*$ and $y \in V^*$, let us define $Z(y \mid x)$ to be $\sum W_0$ if it exists, where W_0 is the set of all plausibility degrees associated with some computation of Z with input x and output y.

When W is W_W, the corresponding machine is called a Turing weighted machine. Similarly, when W is W_X, W_I, W_T, or W_N, the machine is called a Turing maximin machine, a Turing minimax machine, a Turing max-product machine, or a Turing nondeterministic machine, respectively. Also, a Turing weighted machine whose p that is probabilistic will be called a Turing probabilistic machine and any Turing W-machine whose p is deterministic is called a Turing deterministic machine. At this point it is necessary to say again that Turing maximin machines and Turing probabilistic machines operate in essentially the same way as fuzzy algorithms and probabilistic Turing machines operate.

Theorem 4.3.3 *There exists a W-machine with $X \subseteq \Sigma^*$ and $Y \subseteq \Sigma^*$ for some set Σ that for every Turing W-machine $Z = (U, V, S, Q, T, q_0, p)$ with $S \subseteq \Sigma$, there exists a program π that is admissible on $\mathfrak{M}$ with the property that for every $x \in U^*$ and $y \in U^*$, $\mathfrak{M}_\pi(y \mid x)$ and $Z(y \mid x)$ are either both defined or both undefined, and if both are defined, then $\mathfrak{M}_\pi(y \mid x) = Z(y \mid x)$.*

Proof Assume that $\mathfrak{M}$ is a W-machine with $M = (\Sigma \cup \{\#\})^*$, where $\#$ is a special symbol that is not a member of Σ, $X \subseteq \Sigma$, $\Upsilon \subseteq \Sigma$, $\mathfrak{F} = \{MR, ML\} \cup \{F_a \mid a \in \Sigma\}$, and $\mathscr{P} = \{P_a \mid a \in \Sigma\} \cup \mathscr{P}_0$, where for every $x \in X$, $y \in \Sigma$, $\alpha, \beta \in M$, and $t = 1, 2$,

$$\mathfrak{M}_I(\alpha \mid x) = \begin{cases} 1, & \text{if } \alpha = \#x, \\ 0, & \text{otherwise,} \end{cases}$$

$$\mathfrak{M}_O(y \mid \alpha) = \begin{cases} 1, & \text{if } y \text{ is obtained from } \alpha \text{ by omitting } \#, \\ 0, & \text{otherwise,} \end{cases}$$

$$\mathfrak{M}_{F_a}(\beta \mid \alpha) = \begin{cases} 1, & \text{if } \alpha = \gamma\#s\delta, \beta = \gamma\#a\delta \text{ for some } \gamma, \delta \in \Sigma^* \text{ and } s \in \Sigma, \\ 0, & \text{otherwise,} \end{cases}$$

$$\mathfrak{M}_{MR}(\beta \mid \alpha) = \begin{cases} 1, & \text{if } \alpha = \gamma\#s\delta, \beta = \gamma s\#\delta \text{ for some } \gamma, \delta \in \Sigma^* \text{ and } s \in \Sigma, \\ 0, & \text{otherwise,} \end{cases}$$

$$\mathfrak{M}_{ML}(\beta \mid \alpha) = \begin{cases} 1, & \text{if } \alpha = \gamma s\#s'\delta, \beta = \gamma\#ss'\delta \text{ for some } \gamma, \delta \in \Sigma^* \text{ and } s, s' \in \Sigma, \\ 0, & \text{otherwise,} \end{cases}$$

$$\mathfrak{M}_{P_a}(t \mid \alpha) = \begin{cases} 1, & \text{if } t = 1 \text{ and } \alpha = \gamma\#a\delta \text{ for some } \gamma, \delta \in \Sigma^*, \\ & \text{or } t = 2 \text{ and } \alpha \neq \gamma\#a\delta \text{ for all } \gamma, \delta \in \Sigma^*, \\ 0, & \text{otherwise} \end{cases}$$

and $\mathscr{P}_p$ is the collection of all unconditional test functions. Except for the unconditional test functions, machine $\mathscr{M}$ functions just like a machine defined in [117].[4] Suppose that $Q = \{q_0, q_1, \ldots, q_n\}$, $S = \{s_1, s_2, \ldots, s_m\}$, $c_{m+1} = R$, $c_{m+2} = L$, and $c_{m+3} = N$. Then, let us define π to be the program consisting of the following instructions:

$$\begin{aligned}
&\text{Start:} &&\text{goto } L_{01} \\
&L_{kj}: &&\text{if } P_{c_j} \text{ then goto } (L_{k0}^j, L_{kj+1}) \\
&L_{k0}^j: &&\text{if } P_k^j \text{ then goto} \\
& && (L_0^1, L_0^2, \ldots L_0^{m+3}, L_1^1, L_1^2, \ldots L_1^{m+3}, \ldots, L_n^1, L_n^2, \ldots L_n^{m+3}) \\
&L_k^j: &&\text{do } F_{c_j}; \text{ goto } L_{k1} \\
&L_k^{m+1}: &&\text{do } F_{MR}; \text{ goto } L_{k1} \\
&L_k^{m+2}: &&\text{do } F_{ML}; \text{ goto } L_{k1} \\
&L_k^{m+3}: &&\text{halt}
\end{aligned}$$

---

4. The original text mentions that $\mathscr{M}$ functions in essentially the same way as the *Turner* machine defined in [117]. Unfortunately, there is no mention of a Turner machine in [117]. Moreover, to the best of my knowledge, there is no abstract machine called Turner machine! Chances are that this was a typo and that the text refers to the Turing machine mentioned in [117].

where $k = 0, 1, 2, ..., n, j = 1, 2, ..., m$, and $\mathscr{M}_{p_k^j}$ is a W-function from M to $\{1, 2, ..., (n+1)(m+3)\}$ such that

$$\mathfrak{M}_{p_k^j}(\bar{a} \mid \alpha) = \begin{cases} p(q_{\bar{t}} s_\ell \mid s_j, q_i), & \text{if } \alpha = \gamma \# s_j \delta, \bar{a} = \ell + \bar{t}(m + 3), \\ & \bar{t} = 0, 1, ..., n \text{ and } \ell = 1, 2, ..., m + 3, \\ 0, & \text{otherwise.} \end{cases}$$

Now it is easy to verify that $\mathfrak{M}$ and π have the desired properties.   □

Definition 4.3.32 Assume that f is W-function from X to Y. Then, f is computable if and only if there is a Turing W-machine $Z = (U, V, S, Q, T, q_0, p)$ such that $X = U^*$, $Y = V^*$, and $f(y \mid x) = Z(y \mid x)$ for all $x \in X$ and $y \in Y$. In addition, f is *computable* by Z or, equivalently, Z *computes* f.

Definition 4.3.33 A W-function $\mathfrak{M}$ is computable if and only if the output operation, test, and output functions are all computable.

When $\mathfrak{M}$ is computable, then the sets X, Y, and M are finite. In addition, it is not difficult to construct a Turing W-machine so simulate the computations of a program that is admissible on a computable W-machine. More specifically,

Theorem 4.3.4 *Assume that $\mathfrak{M}$ is a computable W-function with input set X and output set Y, and π a program admissible on $\mathfrak{M}$. Then there is a Turing W-machine $Z = (U, V, S, Q, T, q_0, p)$ such that $X = U^*$, $Y = V^*$, and for every $x \in X$ and $y \in Y$, $\mathfrak{M}_\pi(y \mid x)$ and $Z(y \mid x)$ are either both defined or both undefined, and if both are defined they are equal.*

4.3.4   Moraga's Fuzzy Turing Machines

Claudio Moraga [94], borrowing ideas that were put forth by Zadeh and Santos, proposed his own versions of fuzzy Turing machines. First, he proposed a machine with fuzzy states and symbols:

Definition 4.3.34 A fuzzy Turing machine $\mathscr{M}_1$ is a sextuple $(Q, \Sigma, \delta, \sqcup, \triangleright, q_0, H)$, where

- $Q : U_Q \to [0, 1]$ is fuzzy subset of states and U_Q is a finite universe of states;

- $\Sigma : U_\Sigma \to [0, 1]$ is fuzzy subset of symbols, U_Σ is a universe of symbols, and $\# \in U_\Sigma$ is a special symbol such that $\Sigma(\#) = 1$;

- $\sqcup$ is the blank symbol;

- $\triangleright \in \Gamma$ is the *left end symbol*;

- $q_0 \in Q$ is the initial state;

- $H \subseteq Q$ is the set of halting or accepting and rejecting states; and

- $\delta : Q \times \Sigma \rightarrow S(Q) \times S(\Sigma) \times \{L, R, N\}$ is the transition function. Alternatively,

$$\delta : \left(S(Q) \times [0,1]_C \right) \times \left(S(\Sigma) \times [0,1]_C \right) \rightarrow S(Q) \times S(\Sigma) \times \{L, R, N\},$$

where $[0,1]_C \subset [0,1]$ includes only the (classically) computable numbers.

By replacing δ with the following relation

$$\Delta \subset \left(S(Q) \times [0,1]_C \right) \times \left(S(\Sigma) \times [0,1]_C \right) \times S(Q) \times S(\Sigma) \times \{L, R, N\},$$

one obtains a nondeterministic version of the fuzzy Turing machine.

When a fuzzy Turing machine $\mathcal{M}_1$ terminates, the result of its computation is defined as follows:

$$\rho = \left(w, \bigwedge_{i=1}^{n}{}^{*}_{T} \Sigma(w_i) \right) \tag{4.3}$$

where $w \in (U_\Sigma \setminus \{\#\})^*$ denotes a string on the tape, $\bigwedge^*_T$ is the transitive closure of some t-norm $\wedge$,[5] and w_i is the ith symbol of the string w. The equation above associates to the result of a computation a plausibility degree that clearly depends on the membership degrees of the symbols that make up the string. Algorithm 4.3.1 describes how a fuzzy Turing machine computes.

$q \leftarrow q_0;$
while $(q \notin H)$ {
    read s;
    $(q', s') \leftarrow \delta\big((q, Q(q)), (s, \Sigma(s))\big);$
    if $(s' \in \{L, R\})$ {
        move accordingly;
    }
    else {
        $s \leftarrow s'; q \leftarrow q';$
    }
}
compute the plausibility degree using (4.3);

Algorithm 4.3.1: The way machine $\mathcal{M}_1$ computes

It is obvious for Algorithm 4.3.1 that there is nothing special here. In fact, the operation of any machine $\mathcal{M}_1$ can be simulated by an ordinary Turing machine. But, this happens just because everything is "precomputed." However, vagueness is not something that can be "precomputed"; thus, this model is completely unrealistic. In addition, the result is not computed to a degree. Nevertheless, Moraga tried to remedy this drawback by introducing a second form of fuzzy Turing machine:

---

5. A t-norm $\wedge : [0,1] \times [0,1] \rightarrow [0,1]$ can be seen as a fuzzy relation on $[0,1]$; thus, one can compute its transitive closure.

Definition 4.3.35 A fuzzy Turing W-machine $\mathscr{M}_2$ is a sextuple $(\Sigma, Q, q_0, H, \delta, \delta_W)$, where

- Σ is the input alphabet;

- Q is a finite set of states, with $\Sigma \cap Q = \varnothing$;

- $q_0 \in Q$ is the initial state;

- $H \subseteq Q$ is the set of halting or accepting and rejecting states;

- $\delta \subset (Q \times \Sigma) \times (Q \times \Sigma \times \{L, R, N\})$; and

- $\delta_W : (Q \times \Sigma) \times (Q \times \Sigma \times \{L, R, N\}) \to [0,1]_C$ assigns a plausibility degree to each transition of the machine.

Definition 4.3.36 Assume that $C_i \vdash_{\mathscr{M}_1} C_{i+1}$. Then, $\eta_W(C_i, C_{i+1})$ is the degree of *reachability* of C_{i+1} from C_i, and it is defined as follows for all $u, v \in \Sigma^*, a, a' \in \Sigma$, and $q, q' \in Q$:

$$
\eta_W(C_i, C_{i+1} = \begin{cases} \delta_W(q, a, q', a'), & \text{if } C_i = uqav, \quad C_{i+1} = uq'a'v, \\ \delta_W(q, a, q', R), & \text{if } C_i = uqav, \quad C_{i+1} = uaq'v, \\ & \text{or } C_i = uqa, \quad C_{i+1} = uaq'\#, \\ \delta_W(q, a, q', L), & \text{if } C_i = uqav, \quad C_{i+1} = uq'av, \\ & \text{or } C_i = qav, \quad C_{i+1} = q'\#av \\ 0, & \text{otherwise.} \end{cases}
$$

Definition 4.3.37 Assume that $C_0 \vdash_{\mathscr{M}_1}^* C_n$ is a computation from C_0 to C_n. Then, the plausibility degree of this computation is

$$
\Gamma(C_0, C_n) = \eta_W(C_0, C_1) * \eta_W(C_1, C_2) * \cdots * \eta_W(C_{n-1}, C_n),
$$

where $*$ is a computable t-norm.

In the case of a nondeterministic fuzzy Turing W-machine, there are different sequences of configurations that lead from C_0 to C_n, and each sequence has its own plausibility degree. Assume that $D_{0,n}$ is the set of all these different plausibility degrees for the nondeterministic computation $C_0 \vdash_{\mathscr{M}_1}^* C_n$. Then,

$$
\Gamma(C_0, C_n) = \bigvee_{\gamma \in D_{0,n}}^{\star} \gamma, \tag{4.4}
$$

where $\bigvee_T^{\star}$ is the transitive closure of the t-conorm $\star$, which is the dual of the t-norm $*$.

Algorithm 4.3.2 is a precise description of the way machine $\mathscr{M}_2$ operates. Since all steps are completely computable, it turns out this is a machine that offers no new results or ideas.

```
D ← ∅;
for(j = 0; j < 100···0; j++){
    i ← 0; Γ_0 ← 1; q ← q_0;
    while (q ∉ H) {
        (q', s') ← δ(q, s);
        Γ_{i+1} ← τ(Γ_i, δ_W(q, s, q', s'));
        i++; q ← q'; s ← s';
    }
    // the machine stops at C_n
    read result-word and Γ_i;
}
if (Γ_i ∉ D_{0,n} ∪ {Γ_i}) {
    D_{0,n} ← D_{0,n} ∪ {Γ_i};
    Compute Γ(C_0, C_n) with (4.4);
}
```

Algorithm 4.3.2: The way machine $\mathscr{M}_2$ computes

4.3.5   An Extension of Turing W-Machines

Buenaventura Clares and Miguel Delgado [31] introduced a notion that is supposed to be the parallel of recursiveness for fuzzy functions. This work is based on earlier work on W-computability and an extension of Turing W-Machines and their properties. Here I will discuss only the extension of Turing W-Machines. Fuzzy recursion theory is discussed in Sect. 4.6. In what follows, $W = ([0, 1], \oplus, \otimes, <)$ is an ordered semiring.

Definition 4.3.38 A W-Turing machine is *characterized* by the septuple

$$Z = (Q, U, V, S, \delta, \pi, \eta^T),$$

where

- $Q = \{q_1, q_2, ..., q_r\}$ is the finite set of internal states;

- U is the set of input symbols;

- V is the set of output symbols;

- S is the set of tape symbols, $U \cup V \subseteq S$ and $Q \cap S = \emptyset$;

- π is an r-dimensional vector $(\pi_{q_1}, \pi_{q_2}, ..., \pi_{q_r})$, $\pi_{q_i} \in W$, called the *initial designator over states*;

- T is the set of terminal states;

- η^T is an r-dimensional vector, called the *final designator over states*, and it holds that $\eta_i^T = 1$ if $q_i \in T$ and $\eta_i^T = 0$ if $q_i \notin T$; and

- $\delta : (S \times Q) \times \left(Q \times \left(S \cup \{R, L, N\} \right) \right) \to W.$

For all $\alpha, \beta \in D(Z)$ (see Definition 4.3.28 on p. 67), we say that α *precedes* β and write $\alpha \vdash \beta$, if $\mathrm{p}^Z(\beta \mid \alpha) \neq 0$ (see Definition 4.3.29 on p. 68).

Definition 4.3.39 Assume that Z is a W-Turing machine. Then, for every $(u, q) \in U^* \times Q$ and $(q', v) \in Q \times V^*$, a finite sequence $\Lambda = \{\alpha_0, \alpha_1, ..., \alpha_n\}$ of instantaneous descriptions such that

- (i) $\alpha_0 = qu$;

- (ii) $\alpha_n = \xi q' \gamma$ with $v = \xi \gamma$; and

- (iii) $\alpha_i \vdash \alpha_{i+1}, i = 0, 1, 2, ..., n - 1$

is called a *calculus with input (u, q) and output (q', v)* and is written as $\alpha_0 \vdash^* \alpha_n$.

The following expression

$$\omega^\Lambda(q', v \mid u, q) = \mathrm{p}^Z(\alpha_1 \mid \alpha_0) \otimes \mathrm{p}^Z(\alpha_2 \mid \alpha_1) \otimes \cdots \otimes \mathrm{p}^Z(\alpha_n \mid \alpha_{n-1})$$

computes a calculus with a degree.

It is quite possible that different calculi may produce the same output from a particular input. In cases like this, the degree to which a given output (q', v) is yielded by the input (u, q) is given by

$$\omega(q', v \mid u, q) = \bigoplus_{\Lambda \subset D(Z)}^* \omega^\Lambda(q', v \mid u, q),$$

where $\oplus^*$ is the extension of the operator $\oplus$ for the set of calculi with input (u, q) and output (q', v). When this set is empty, then the result is undefined.

For some machine Z, we define the plausibility degree to which Z starts with an initial designator of states π and the word $u \in U^*$ printed on the tape and terminates with the word $v \in V^*$ printed on the tape as follows:

$$p_\pi(v \mid u) = \bigoplus_{i=1}^r {}^* \left\{ \pi_{q_i} \otimes \bigoplus_{j=1}^r {}^* \left[\omega(q_j, v \mid u, q_i) \otimes \eta_{q_j}^T \right] \right\}.$$

Definition 4.3.40 A word $u \in U^*$ is W-computable if there is at least one output word $v \in V^*$ such that $p_\pi(v \mid u) \neq 0$.

It is not difficult to extend these notions to functions with more than one variable.

Definition 4.3.41 Suppose that U and V are finite alphabets. Then, a function $f : (U^*)^k \times V^* \to W, k \geq 1$, is partially W-computable if there exists a set $T \subseteq (U^*)^k$ and a W-Turing machine Z such that

$$f(v \mid u^k) = p^k(v \mid u^k), \quad \forall u^k \in T, v \in V^*.$$

If $T = (U^*)^k$, then f is W-computable.

When $T = (U^*)^k$, the machine computes at least an output $v \in V^*$ with plausibility degree $p^k(v \mid u^k) \in W$ in a finite number of steps for any input $u^k \in (U^*)^k$. Otherwise, there are inputs that can make the machine run for ever.

4.4   The Computational Power of Fuzzy Turing Machines

Jiří Wiedermann [142] proposed a nondeterministic fuzzy Turing machine with hypercomputational capabilities. In a nutshell, Wiedermann had shown that his machine can solve problems no ordinary Turing machine can solve. Let me start by giving the definition of his conceptual computing device:

Definition 4.4.1 A nondeterministic fuzzy Turing machine with a unidirectional tape is a nonuple

$$\mathscr{F} = (Q, T, I, \Delta, \sqcup, q_0, q_f, \mu, *),$$

where:

- Q is a finite set of states;

- T is a finite set of tape symbols;

- I is a set of input symbols, where $I \subseteq T$; and

- Δ is a transition relation and it is a subset of $Q \times T \times Q \times T \times \{L, N, R\}$. Each action that the machines takes is associated with an element $\delta \in \Delta$. In particular, for $\delta = (q_i, t_i, q_{i+1}, t_{i+1}, d)$ this means that when the machine is in state q_i and the current symbol that has been read is t_i, then the machine will enter state q_{i+1}, the symbol t_{i+1} will be printed on the current cell, and the scanning head will move according to the value of d, that is, if d is L, N, or R, then the head will move one cell to the left, will not move, or will move one cell to the right, respectively.

- $\sqcup \in T \setminus I$ is the blank symbol;

- q_0 and q_f are the initial and the final state, respectively;

- $\mu : \Delta \to [0,1]$ is a fuzzy relation on Δ; and

- $*$ is a t-norm.

Definition 4.4.2 When μ is partial function from $Q \times T$ to $Q \times T \times \{L, N, R\}$ and T is a fuzzy subset of Q, then the resulting machine is called a deterministic fuzzy Turing machine.

A configuration gives the position of the scanning head, of what is printed on the tape, and the current state of the machine. If S_i and S_{i+1} are two configurations, then $S_i \vdash^\alpha S_{i+1}$ means that S_{i+1} is reachable in one step from S_i with a plausibility degree that is equal to α if and only if there is a $\delta \in \Delta$ such that $\mu(\delta) = \alpha$ and by which the machine goes from S_i to S_{i+1}. When a machine starts with input some string w, the characters of the string are printed on the tape starting from the leftmost cell; the scanning head is placed atop the leftmost cell, and the machine enters state q_0. If

$$S_0 \vdash^{\alpha_0} S_1 \vdash^{\alpha_1} S_2 \vdash^{\alpha_2} \cdots \vdash^{\alpha_{n-1}} S_n,$$

then S_n is reachable from S_0 in n steps. Assume that S_n is reachable from S_0 in n steps, then the plausibility degree of this *computational path* is

$$D\big((S_0, S_1, ..., S_n)\big) = \begin{cases} 1, & n = 0, \\ D\big((S_0, S_1, ..., S_{n-1})\big) * \alpha_{n-1}, & n > 0. \end{cases}$$

Obviously, the value that is computed with this formula depends on the specific path that is chosen. Since the machine is nondeterministic, it is quite possible that some configuration S_n can be reached via different computational paths. Therefore, when a machine starts from S_0 and finishes at S_n in n steps, the plausibility degree of this computational path, which is called a *computation*, should be equal to the maximum of all possible computation paths:

$$d(S_n) = \max\Big[D\big((S_0, S_1, ..., S_n)\big)\Big].$$

In different words, the plausibility degree of the computation is equal to the plausibility degree of the computational path that is most likely to happen.

Assume that a machine starts from configuration S_0 with input the string w. Then, a computational path $S_0, S_1, ..., S_m$ is an accepting path of configurations if the state of S_m is q_f. In addition, the string w is accepted with degree equal to $d(S_m)$.

Definition 4.4.3 Assume that $\mathscr{F}$ is a fuzzy nondeterministic Turing machine. Then, an input string w is accepted with plausibility degree $e(w)$ by $\mathscr{F}$ if and only if

- there is an accepting configuration from the initial configuration S_0 on input w; and

- $e(w) = \max_S \{d(S) \mid S$ is an accepting configuration reachable from $S_0\}$.

Also,

Definition 4.4.4 The fuzzy language accepted by some machine $\mathscr{F}$ is the fuzzy set that is defined as follows:

$$L(\mathscr{F}) = \Big\{\big(w, e(w)\big) \,\Big|\, w \text{ is accepted by } \mathscr{F} \text{ with plausibility degree } e(w)\Big\}.$$

The class of all fuzzy languages accepted by a fuzzy Turing machine, in the sense just explained, with (classically) computable t-norms is denoted as Φ.

Theorem 4.4.1 $\Phi = \Sigma_1^0 \cup \Pi_1^0$, *that is, Φ is the union of the recursively enumerable and the co-recursively enumerable languages.*

Proof (Sketch) In order to show that $\Phi = \Sigma_1^0 \cup \Pi_1^0$ one has to prove that $\Sigma_1^0 \cup \Pi_1^0 \subseteq \Phi$ and, at the same time, that $\Phi \subseteq \Sigma_1^0 \cup \Pi_1^0$. Assume that L is a language such that $L \in \Sigma_1^0 \cup \Pi_1^0$. Then either $L \in \Sigma_1^0$ or $L \in \Pi_1^0$. When L is recursively enumerable, there is a nonfuzzy machine $\mathcal{M}$ that is able to semidecide L. In what follows, it will be shown that for each constant $0 \leq c < 1$ there is a fuzzy Turing machine $\mathcal{F}$, whose acceptance criterion is given in Definition 4.4.4, such that $w \in L$ if and only if w is accepted by $\mathcal{F}$ with plausibility degree equal to 1, that is, $(w, 1) \in L(\mathcal{F})$, and $w \notin L$ if and only if $(w, c) \in L(\mathcal{F})$. Given any t-norm and $0 \leq c < 1$, one can specify $\mathcal{F}$ as follows: unless it is explicitly stated, by default all commands will have plausibility degree that is equal to one. Suppose that w is the input to $\mathcal{F}$. Then, this machine will make a nondeterministic branch and one path will lead to the simulation of $\mathcal{M}$. In addition, when $\mathcal{M}$ enters an accepting state, $\mathcal{F}$ will enter an accepting state q with plausibility degree equal to one. Also, the other path leads to an accepting state q' with plausibility degree equal to c. Now, if $w \in L$, then both q and q' are reached. In addition, w will be accepted in q with plausibility degree equal to one (i.e., $(w, 1) \in L(\mathcal{F})$). If $w \notin L(\mathcal{F})$, then the machine will not enter any accepting state but r and so $(w, c) \in L(\mathcal{F})$. Assume now that L is co-recursively enumerable and so $\overline{L}$ is recursively enumerable. This implies that there is a machine $\mathcal{M}'$ that recognizes $\overline{L}$. Following an argument similar to the one presented so far, one can show that for each constant $0 \leq c < 1$ there is a fuzzy Turing machine $\mathcal{G}$ such that $w \in \overline{L}$ if and only if w is accepted by $\mathcal{G}$ with plausibility degree equal to one (i.e., $(w, 1) \in L(\mathcal{G})$) and $w \notin \overline{L}$ if and only if $(w, c) \in L(\mathcal{G})$. In different words, $w \notin L$ if and only if $(w, 1) \in L(\mathcal{G})$ and $w \in L$ if and only if $(w, c) \in L(\mathcal{G})$. And this concludes the first part of the proof. Let me now proceed with the second part.

Now it will be shown that for any fuzzy Turing machine $\mathcal{F} = (Q, T, I, \Delta, \llcorner, q_0, q_f, \mu, *)$ with a computable t-norm there are crisp languages $L_1 \in \Sigma_1^0$ and $L_2 \in \Pi_1^0$ such that $(w, d) \in L(\mathcal{F})$ if and only if $w\#d \in L_1 \cap \in L_2 \Sigma_1^0 \cup \Pi_1^0$ for any $w \in \left(I \setminus \{\#\}\right)^*$ and $\# \in T$. Assume that F is fuzzy Turing machine where all commands have plausibility degrees that are equal to one (essentially, this machine is a nondeterministic Turing machine). Also, assume that $ACF_{\mathcal{F}}(w)$ is a set that contains all accepting configurations of $\mathcal{F}$ on input w and that e is the evaluation function that assigns to each accepting configuration its plausibility degree. In addition, consider the commutative ordered semigroup $G = ([0, 1], *, \leq)$, where $*$ is a t-norm and $\leq$ the usual ordering relation. Furthermore, let $\mathbb{Q} \supset J = \{\alpha_1, \alpha_2, ..., \alpha_k\} \subset [0, 1]$, where $\alpha_1 < \alpha_2 < \cdots < \alpha_k$, be the range of μ and $G(J)$ be a subsemigroup of G generated by replacing $[0, 1]$ with a set that has as elements all elements of J and all elements produced by the t-norm, that is, when α_m and α_m are elements of this set, then $\alpha_m * \alpha_n$ belongs to this set also. It is relatively easy to see that

$$\Sigma_1^0 \ni L_1 = \left\{w\#d \mid \exists a \in ACF_{\mathcal{F}}(w) : e(a) = d \in G(J)\right\},$$
$$\Pi_1^0 \ni L_2 = \left\{w\#d \mid \forall a \in ACF_{\mathcal{F}}(w) : e(a) \leq d \in G(J)\right\}.$$

In particular, in order to show that $L_1 \in \Sigma_1^0$, one should consider an ordinary nondeterministic machine $\mathcal{M}_1$, which on input $w\#d$, where $d = \alpha_n * \alpha_m$, first guesses the numbers α_n and α_m, then computes $\alpha_n * \alpha_m$, and checks whether $d \in G(J)$. Next, $\mathcal{M}_1$ guesses a and simulates F on w to see whether $a \in ACF_{\mathcal{F}}(w)$ and $e(a) = d$. In order to show that $L_2 \in \Pi_1^0$, one should consider an ordinary nondeterministic machine M_2 that accepts the language

$$\overline{L}_2 = \left\{ w\#d \mid \exists a \in ACF_{\mathscr{F}}(w) : e(a) > d \right\}.$$

By a similar argument one concludes that $\overline{L}_2 \in \Sigma_1^0$ and so $L_2 \in \Pi_1^0$. Furthermore, it is clear that $(w, d) \in L(\mathscr{F})$ if and only if $w\#d \in L_1 \cap L_2$ because the second condition is just a reformulation of conditions for $\mathscr{F}$ to accept w with plausibility degree equal to d according to Definition 4.4.4.   □

Benjamín Callejas Bedregal and Santiago Figueira [9] have criticized the work done by Wiedermann—they claim that because Wiedermann employs a certain fuzzification process, his principal results are not valid. In particular, they consider a special way to transform crisp sets into fuzzy sets. For any language A and rational numbers $0 \le a, b \le 1$, one can define the following fuzzification of A:

$$F_A(a, b) = \left\{ (w, a) \mid w \in A \right\} \cup \left\{ (w, b) \mid w \notin A \right\}.$$

The first part of Theorem 4.4.1 can be restated as follows:

Theorem 4.4.2 *Assume that $A \subseteq \Sigma^*$ and $0 \le a < 1$.[6] Then, if A is recursively enumerable or co-recursively enumerable, then $F_A(a, 1)$ or $F_A(1, a)$, respectively, is a language accepted by some fuzzy Turing machine.*

In addition, these authors prove the following stronger result:

Theorem 4.4.3 *Suppose that $A \subseteq \Sigma^*$ and $a, b \in \mathbb{Q}$, where $0 \le b < a \le 1$. Then the following are equivalent:*

(i) *A is recursively enumerable; and*

(ii) *there is fuzzy Turing machine in C [i.e., the class of all fuzzy Turing machines with rational plausibility degrees and (classically) computable t-norms] that accepts the language $F_A(a, b)$.*

Next, they claim that Wiedermann is using two different methods to fuzzify recursively enumerable and co-recursively enumerable sets. However, the truth is that when $L \in \Pi_1^0$, he considers the language $\overline{L} \in \Sigma_1^0$ and so he does not employ a different fuzzification method. In addition, it is claimed that in the proof of the second part of Theorem 4.4.1

...all it shows is that there exist r.[ecursively]e.[enumerable] languages L_1 and L_2 such that $L_1 \setminus L_2 = \left\{ w\#d \mid (w, d) \in L(\mathscr{F}) \right\}$ for every fuzzy Turing machine $\mathscr{F}$.

Again the problem is that Wiedermann clearly claims that $(w, d) \in L(\mathscr{F})$ if and only if $w\#d \in L_1 \cap L_2$. All in all, I am not really convinced that the arguments of Bedregal and Figueira disprove Wiedermann's result.

---

6. In the original paper, the second condition is "$0 \le b < 1$," which is obviously a typo since in what follows the authors use "a" instead of "b".

4.5   On Universal Fuzzy Turing Machines

On p. 44 of [18], the famous book on computability and logic by George S. Boolos and Richard C. Jeffrey, there is an exercise that asks the reader to "[s]how that if Turing's thesis [i.e., the Church–Turing thesis] is correct, then a universal Turing machine must exist." But what if the Church–Turing thesis is not correct? Does this imply that there is no universal Turing machine? Yes it does! In fact, Selim G. Akl [1, 2] has argued that there is no universal computer at all! Provided that Akl's argument has no flaws, one can conclude that there is no universal Turing machine either. In a nutshell, according to Akl, only a machine that can perform an infinite number of operations per step can be classified as a universal machine, in different words only a machine that is able to perform a *supertask* (see [126] for a discussion of supertasks and their feasibility). Now, if supertasks are impossible, then, according to Akl, there are no universal machines and, consequently, there is no universal fuzzy Turing machine either. Interestingly, it has been shown that there is no universal fuzzy Turing machine, but this does not prove that supertasks are impossible...

To the best of my knowledge, Yong Ming Li [87] was the first researcher who argued against the existence of a universal fuzzy Turing machine. Later on, Li [84, 85] just elaborated his work without presenting any new results. Furthermore, Li [86] extended his work to L-Turing machines (i.e., vague machines where the plausibility degrees are drawn from some lattice L), which is essentially a reformulation of his earlier work. Before presenting his argument, we need a few auxiliary definitions.

Assume that $\lambda : \Sigma^* \to [0, 1]$ is a fuzzy language. Then, when its support, $S(\lambda)$, is finite, it makes sense to claim that λ can be represented by

$$\left\{ \left(s, \lambda(s) \right) \,\middle|\, (s \in \Sigma^*) \wedge \left(\lambda(s) \neq 0 \right) \right\}.$$

Also, $^{[\alpha]}A$ is defined as follows:

$$^{[\alpha]}A = \left\{ x \,\middle|\, \left(x \in X \right) \wedge \left(A(x) = \alpha \right) \right\}.$$

Assume that $\mathcal{M}$ is a fuzzy Turing machine. Then, $\lambda_{\mathcal{M}} : \Sigma^* \to [0, 1]$ is a function that denotes the fuzzy language accepted by $\mathcal{M}$. This function is precisely defined in Li's papers as follows:

$$\lambda_{\mathcal{M}}(s) = \bigvee \left\{ \left(\vdash^* (q_0 s, \alpha_1 q \alpha_2) \right) * F(q) \,\middle|\, q \in Q, \alpha_1, \alpha_2 \in T^* \right\},$$

where $\vdash^*$ is the reflexive and transitive closure of $\vdash$ (i.e., the fuzzy step relation), $F(q)$ is plausibility degree that the machine will enter state q, and all others are as in Wiedermann's definition. In different words, the set of accepting states is a fuzzy subset of Q in Li's definition, while it is a crisp subset in Wiedermann's definition.

Assume that $\lambda : \Sigma^* \to [0, 1]$ is a fuzzy language. Then, when there is a nondeterministic fuzzy Turing machine such that $\lambda = \lambda_{\mathcal{M}}$, λ is a *fuzzily recursively enumerable* language. In addition, if for any any $s \in \Sigma^*$ machine $\mathcal{M}$ halts, λ is a *fuzzily recursive* language that can be decided by $\mathcal{M}$.

Theorem 4.5.1 *Suppose that $\lambda : \Sigma^* \to [0, 1]$ is a fuzzy language. Then, the following three statements are equivalent:*

(i) λ is decided by a nondeterministic fuzzy Turing machine whose t-norm is the min function;

(ii) $S(\lambda)$ is finite and for any $a \in S(\lambda)$, $^a\lambda$ is recursively enumerable; and

(iii) λ is fuzzily recursively enumerable.

Proof (i)⇒(ii) Assume that $\lambda = \lambda_{\mathcal{M}}$, where $\mathcal{M}$ is nondeterministic fuzzy Turing machine whose t-norm is the min function. Thus,

$$\lambda(s) = \bigvee \left\{ \min\{\vdash^* (q_0 s, \alpha_1 q \alpha_2), F(q)\} \,\Big|\, q \in Q, \alpha_1, \alpha_2 \in T^* \right\}.$$

From the definition of the step relation (see Sect. 2.2.1) it follows that $S(\lambda) \subseteq \{0, 1\} \cup S(\delta) \cup S(F)$, which implies that $S(\lambda)$ is finite. In addition, for any $a \in S(\lambda)$ and $\Sigma^* \ni s \in {}^a\lambda$, $\lambda(s) \geq a$ which is equivalent to $\exists q \in Q, \min\{\vdash^* (q_0 s, \alpha_1 q \alpha_2), F(q)\} \geq a$. In turn, this means that $\vdash^* (q_0 s, \alpha_1 q \alpha_2) \geq a$ and $F(q) \geq a$ and so $s \in L(\mathcal{M}_a)$, where $\mathcal{M}_a$ is nondeterministic Turing machine, that is, $^a\lambda$ is decided by $\mathcal{M}_a$ and thus $^a\lambda$ is recursively enumerable.

(ii)⇒(iii) Since $^a\lambda$ is recursively enumerable for any $a \in S(\lambda)$, it can be decided by nondeterministic Turing machine $\mathcal{M}$. Define $a \wedge {}^a\lambda$ to be the fuzzy subset of Σ^*:

$$(a \wedge {}^a\lambda)(s) = \begin{cases} a, & \text{when } s \in \lambda_a, \\ 0, & \text{otherwise.} \end{cases}$$

Now, $a \wedge {}^a\lambda$ can be decided by nondeterministic Turing machine and thus it is fuzzy recursively enumerable. Also, $S(\lambda)$ is finite and the family of fuzzily recursively enumerable languages is closed under finite union operations, since the union of two fuzzily recursive/recursively enumerable languages is still fuzzily recursive/recursively enumerable. Thus,

$$\lambda = \bigvee_{a \in S(\lambda)} (a \wedge {}^a\lambda)$$

is fuzzily recursively enumerable, that is, λ can be decided by a nondeterministic fuzzy Turing machine.

(iii)⇒(i) This is obvious a nondeterministic fuzzy Turing machine that uses min as its t-norm is a special case of nondeterministic fuzzy Turing machines that may use any t-norm. □

Also, let us assume without loss of generality that all plausibility degrees involved in Li's argument are rational numbers. The following results are important for the final proof:

Theorem 4.5.2 *Assume that $\lambda : \Sigma^* \to [0, 1]$ is a fuzzy language. Then, the following statements are equivalent:*

(i) λ can be decided by a deterministic fuzzy Turing machine; and

(ii) $S(\lambda)$ is finite and $^{[a]}\lambda$ is recursively enumerable for any $a \in S(\lambda)$.

Proof (i)⇒(ii) Assume that $\lambda = \lambda_{\mathcal{M}}$, where $\mathcal{M}$ is a deterministic fuzzy Turing machine. In addition, $\lambda(s) = F(q)$ when $q_0 s \vdash^* \alpha_1 q \alpha_2$ for some accepting state q, that is, $F(q) > 0$ and $\lambda_{\mathcal{M}}(s) = 0$ for all other cases. Also, it holds that $S(\lambda) \subseteq S(F) \cup \{1\}$, which implies that $S(\lambda)$ is finite. Furthermore, whenever $a \in S(\lambda)$, then $s \in {}^{[a]}\lambda$ if and only if $\lambda(s) = a$. This equality is equivalent to $\exists q \in Q, q_0 s \vdash^* \alpha_1 q \alpha_2$, and $F(q) = a$. Again, this is equivalent to $s \in L(\mathcal{M}_{[a]})$, where $\mathcal{M}_{[a]}$ is a deterministic fuzzy Turing machine, as defined in Definition 4.4.2. This implies that ${}^{[a]}\lambda = L(\mathcal{M}_{[a]})$ is recursively enumerable for any $a \in S(\lambda)$.

(ii)⇒(i) It is obvious that the following holds true:

$$\lambda = \bigvee_{a \in S(\lambda)} (a \wedge {}^{[a]}\lambda).$$

Since ${}^{[a]}\lambda$ is recursively enumerable for any $a \in S(\lambda)$, this means that ${}^{[a]}\lambda$ can be decided by a deterministic Turing machine. In addition, $a \wedge {}^{[a]}\lambda$ can be decided by a deterministic fuzzy Turing machine $\mathcal{M}'$. Also, if a and b are two distinct elements of $R(\lambda)$, then ${}^{[a]}\lambda \cap {}^{[b]}\lambda = \varnothing$ and thus $S(a \wedge {}^{[a]}\lambda) \cap S(b \wedge {}^{[b]}\lambda) = {}^{[a]}\lambda \cap {}^{[b]}\lambda = \varnothing$. Now, it can be proved that when two fuzzy languages λ_1 and λ_2, such that $S(\lambda_1) \cap S(\lambda_2) = \varnothing$, can be decided by two deterministic fuzzy Turing machines, then their union can also be decided by a deterministic fuzzy Turing machine. So, this result and the fact that $R(\lambda)$ is finite leads to the conclusion that $\lambda = \bigvee_{a \in S(\lambda)} (a \wedge {}^{[a]}\lambda)$ can be decided by a deterministic fuzzy Turing machine. □

It is rather important to stress that this result does not hold true for fuzzily recursively enumerable languages.

Theorem 4.5.3 *Assume that $\lambda : \Sigma^* \to [0, 1]$ is a fuzzy language. Then, the following statements are equivalent:*

(i) λ can be decided by a nondeterministic fuzzy Turing machine;

(ii) $S(\lambda)$ is finite and ${}^a\lambda$ is recursive for any $a \in S(\lambda)$;

(iii) $S(\lambda)$ is finite and ${}^{[a]}\lambda$ is recursive for any $a \in S(\lambda)$; and

(iv) λ can be decided by a deterministic fuzzy Turing machine.

Proof The proofs of Theorems 4.5.1 and 4.5.2 can be used to prove the equivalence between clauses (i) and (ii), and (iii) and (iv), respectively. The implication (iv)⇒(i) is obvious since a deterministic fuzzy Turing machine is actually a special case of a nondeterministic machine. The implication (ii)⇒(iii) holds true because of the following equality:

$$^{[a]}\lambda = {}^a\lambda - \bigcup \left\{ {}^b\lambda \mid b \in S(\lambda), b > a \right\}$$

for any $a \in S(\lambda)$, and the family of recursive languages is closed under finite union and complementation. □

Assume that a universal fuzzy Turing machine exists. Then if $\mathcal{U}$ is such a machine, it will have as input code numbers that correspond to the controlling device of an ordinary fuzzy Turing machine as well as its input. In practice, it is convenient to consider a machine that has as input an integer that corresponds to both the controlling device and the input. The machine will be able to compute the plausibility degree to which the input would be accepted by an ordinary fuzzy machine. In particular, $\lambda_{\mathcal{U}}(\langle\langle\mathcal{M}\rangle,s\rangle) = \lambda_{\mathcal{M}}(s)$, where $\langle\langle\mathcal{M}\rangle,s\rangle$ is the code number of both the machine $\mathcal{M}$ and its input s. In order to keep things simple, it is considered that $\Sigma = \{0,1\}$. Moreover, $\lambda_{\mathcal{U}}(\langle\langle\mathcal{M}\rangle,s\rangle) \in \mathbb{Q}\cap(0,1]$. Function $\lambda_{\mathcal{U}}$ can be "defined" as follows:

$$\lambda_{\mathcal{U}}(\theta) = \begin{cases} q, & \theta = \langle\langle\mathcal{M}\rangle,s\rangle \text{ and } \lambda_{\mathcal{M}}(s) = q, \\ 0, & \text{otherwise.} \end{cases} \tag{4.5}$$

In addition, the following result gives us information about the image set of $\lambda_{\mathcal{U}}$:

Theorem 4.5.4 $\mathbb{Q}\cap(0,1] \subseteq S(\lambda_{\mathcal{U}})$, where $S(\lambda_{\mathcal{U}})$ includes all values of $\lambda_{\mathcal{U}}$ that are greater than zero.

Proof Consider a sequence of n nondeterministic fuzzy Turing machines that may have any t-norm. For each such machine $\mathcal{M}$ it should hold true that $\lambda_{\mathcal{M}}(s) = k/n$ for some $s \in \Sigma^*$ and any $k \leq n$. Given the integers k and n, one can define a fuzzy language $\lambda_n : \{0,1\}^* \to [0,1]$ as follows:

$$\lambda_n(s) = \begin{cases} k/n, & \text{when } s = a^k b^k, k \leq n, \\ 0, & \text{otherwise.} \end{cases}$$

Examples of such languages include the following ones:

$$\lambda_2 = \left\{(ab,1/2),(a^2b^2,1)\right\}$$
$$\lambda_3 = \left\{(ab,1/3),(a^2b^2,2/3),(a^3b^3,1)\right\}$$
$$\cdots$$
$$\lambda_n = \left\{(ab,1/n),(a^2b^2,2/n),\dots,(a^nb^n,1)\right\}.$$

The support of λ_n is finite and so can easily deduce that this language is a fuzzy regular language, thus a fuzzily recursively enumerable language. Therefore, there exists a nondeterministic fuzzy Turing machine $\mathcal{M}_n$ such that $\lambda_{\mathcal{M}_n} = \lambda_n$. In addition, from the construction of $\lambda_{\mathcal{U}}$ in (4.5), when $\theta = \langle\langle\mathcal{M}_n\rangle,a^kb^k\rangle$ and since $\lambda_{\mathcal{M}_n}(a^kb^k) = k/n$, then it is known that $\lambda_{\mathcal{U}}(\theta) = k/n$ for any positive integers n and k, such that $k \leq n$. All these imply that $\mathbb{Q}\cap(0,1] \subseteq S(\lambda_{\mathcal{U}})$. $\square$

Assume that $*$ is a t-norm. Then, for any $a \in [0,1]$, one can inductively define the power operation of a as follows: $a^0 = 1, a^1 = a$, and $a^{n+1} = a^n * a$. For any $D \subseteq [0,1]$, the subalgebra of $([0,1],*)$ generated by D is defined as follows:

$$\text{Sub}(D) = \left\{a_1^{l_1} * \cdots * a_k^{l_k} \,\middle|\, a_1,\dots,a_k \in D \text{ and } l_1,\dots,l_k \in \mathbb{N}\right\}.$$

Lemma 4.5.1 *Assume that ∗ is a t-norm and D a finite subset of* $\mathbb{Q} \cap (0, 1]$. *Then,* $\mathbb{Q} \cap (0, 1] \not\subseteq$ Sub(D).

Proof If $D = \{1\}$, then obviously $\mathbb{Q} \cap (0, 1] \not\subseteq$ Sub(D), since Sub(D) = $\{1\}$. Suppose that D contains some elements that belong to $\mathbb{Q} \cap (0, 1]$ but are different from the number 1. Let $D \setminus \{1\} = \{a_1, \cdots, a_k\}$ such that $a_1 < \cdots < a_k < 1$. For any $a \in$ Sub(D), when $a \neq 1$, then $a = a_1^{l_1} * \cdots * a_k^{l_k}$ for some nonnegative integers $l_1, \ldots, l_k$ such that not all numbers are equal to zero. Also, $a \leq a_k$ because $x * y \leq \min\{x, y\}$ for all $x, y \in [0, 1]$. This means that a_k is the largest element of the set Sup(D) $\setminus \{1\}$, and thus $(a_k, 1) \cap$ Sup(D) = ∅. However, $(a_k, 1) \cap \mathbb{Q} \cap (0, 1] = (a_k, 1) \cap \mathbb{Q} \neq$ ∅. Take $x \in (a_k, 1) \cap \mathbb{Q}$, then $x \notin$ Sup(D) and hence $\mathbb{Q} \cap (0, 1] \not\subseteq$ Sub(D). □

It follows from Theorem 4.5.4 that $\mathbb{Q} \cap (0, 1] \subseteq S(\lambda_{\mathscr{U}})$. Now, for any nondeterministic Turing machine $\mathscr{M}$, there exists a finite set $D \subseteq \mathbb{Q} \cap (0, 1]$ such that $S(\lambda_{\mathscr{M}}) \subseteq$ Sub(D). Lemma 4.5.1 allows one to conclude that $S(\lambda_{\mathscr{M}}) \neq S(\lambda_{\mathscr{U}})$, which means that $\lambda_{\mathscr{U}}$ cannot be decided by any nondeterministic fuzzy Turing machine. Therefore, one can safely conclude the following:

Corollary 4.5.1 *In general, a universal fuzzy Turing machine does not exist.*

Fuzzy set theory is about imprecision and inexactness; therefore, one would reasonably argue that it makes no sense to expect to devise a machine that can fully simulate any fuzzy Turing machine. On the other hand, it is more reasonable to expect the "construction" of a machine that can approximately simulate any fuzzy Turing machine. Indeed, Li [84] has proposed such a machine as a restricted form of a universal fuzzy Turing machine. Obviously, in the lights of Akl's work, even this restricted conceptual computing device does not count for a universal machine. Nevertheless, it makes sense to present the construction for reasons of completeness, if for nothing else.

If the plausibility degrees of the transition functions of all nondeterministic fuzzy Turing machines are restricted to a finite subset of $[0, 1]$, then it is possible to construct nondeterministic fuzzy machine to recognize $\lambda_{\mathscr{U}}$. For example, any set $D \subseteq \mathbb{Q} \cap [0, 1]$ such that $0, 1 \in D$ can be used for this purpose. Let now see how one can encode any nondeterministic fuzzy machine as bitstrings.

Assume that $M = (Q, \{0, 1\}, \Gamma, \delta, q_1, \llcorner, \{q_2\})$ is nondeterministic fuzzy Turing machine. Also, $Q = \{q_1, q_2, \ldots, q_n\}$, where q_1 is the initial state and q_2 is the single accepting state, and $\Gamma = \{X_1, X_2, \ldots, X_m\}$, where $m \geq 3$ and X_1 is the symbol 0, X_2 is the symbol 1, X_3 is the symbol $\llcorner$, etc. In addition, the symbols D_1 and D_2 will denote the directions L and R, respectively. Then, a transition $(q_i, X_j, q_k, X_l, D_s, a_t) \in \delta$ will be encoded by a bitstring of the following form:

$$0^i 10^j 10^k 10^l 10^s 10^t.$$

Now, it is possible to encode the whole machine with a bitstring of the following form:

$$111\text{code}_1 1\text{code}_2 11 \ldots 11\text{code}_p 111,$$

where each code$_i$ corresponds to some encoded transition. Clearly, one has to somehow order the transitions but is a minor problem. The encoding of some machine M will be denoted as $\langle M \rangle$. Provided that s is the input of some machine M, the concatenation of $\langle M \rangle$ and s is $\langle \langle M \rangle, s \rangle$. Let us now redefine function $\lambda_{\mathscr{U}}$:

$$\lambda_{\mathscr{U}}(\theta) = \begin{cases} a, & \theta = \langle \langle M \rangle, s \rangle \text{ and } \lambda_{\mathscr{M}}(\theta) = a, \\ 0, & \text{otherwise.} \end{cases}$$

Theorem 4.5.5 *By restricting the plausibility degrees of the transition function of some nondeterministic fuzzy Turing machine to elements of some set $D \subseteq \mathbb{Q} \cap [0,1]$ such that $0, 1 \in D$, the language $\lambda_{\mathscr{U}}$ can be accepted by this restricted machine.*

4.6   Fuzzily Recursive Sets

The notions of fuzzily recursive and fuzzily recursively enumerable sets were introduced by Leon Harkleroad [63]. In particular, Harkleroad considered only fuzzy subsets of $\mathbb{N}$ in order to "remain within the framework of classical recursion theory." For reasons of brevity, in what follows, $\bar{I}$ will stand for $\mathbb{Q} \cap [0,1]$.

Definition 4.6.1 A fuzzy subset $S : \mathbb{N} \to \bar{I}$ is *fuzzily recursive* if and only if function S is a recursive function. Also, S is *fuzzily recursively enumerable* if and only if the support of S is a partial function that is recursive.

This definition is not general enough since most fuzzy subsets are not subsets of $\mathbb{N}$. Loredana Biacino and Giangiacomo Gerla have extended the notion of fuzzily recursive and fuzzily recursively enumerable sets to include fuzzy subsets of any ordinary set. Although this work was presented in a series of papers, Gerla's [49] monograph has been mostly used for the exposition that follows.

The following result brings forth an alternative formulation of recursive enumerability:

Proposition 4.6.1 *A subset A of X is recursively enumerable if and only if there is a recursive function $h : X \times \mathbb{N} \to \{0, 1\}$, which is increasing with respect to the second variable, such that for all $a \in X$:*

$$\chi_A(a) = \lim_{n \to +\infty} h(a, n), \tag{4.6}$$

where $\lim$ is the symbol for the limit operator.

Proof Suppose that A is recursive enumerable. Then, when $A = \varnothing$, (4.6) holds true by simply letting h be the function that is constantly equal to zero. Otherwise, suppose that $f : \mathbb{N} \to X$ is a recursive function whose range is A (see Definition 2.5.2 on p. 36) and define h as follows:

$$h(x, n) = \begin{cases} 1, & \text{when } x \in \{f(1), f(2), ..., f(n)\}, \\ 0, & \text{otherwise.} \end{cases}$$

Clearly, function h is recursive, increasing with respect to the second variable, and (4.6) holds true.

Assume now that (4.6) holds true and that h is a recursive function that is increasing with respect to the second variable. Then, the following function

$$f(x) = \begin{cases} 1, & \text{when there is an } n \text{ such that } h(x, n) > 0, \\ \perp, & \text{otherwise} \end{cases}$$

is a partial recursive function whose domain is A, which implies that A is recursively enumerable. □

Using this result, one can extend the notion of fuzzily recursively enumerable sets as follows:

Definition 4.6.2 A fuzzy subset $A : X \rightarrow [0, 1]$ is fuzzily recursively enumerable when there is a recursive function $h : X \times \mathbb{N} \rightarrow \bar{I}$, such that for every $x \in X$, $h(x, n)$ is increasing with respect to n and

$$A(x) = \lim_{n \to +\infty} h(x, n). \tag{4.7}$$

Remark 4.6.1 The previous equation is equivalent to the following one:

$$A(x) = \bigvee \{h(x, n) \mid n \in \mathbb{N}\},$$

because h is increasing with respect to n.

Proposition 4.6.2 *Assume that X is a crisp set. Then, for every fuzzy subset A of X the following are equivalent:*

(i) A is fuzzily recursively enumerable;

(ii) the set $K(A) = \left\{(x, \lambda) \mid \left((x, \lambda) \in X \times \bar{I}\right) \wedge \left(A(x) > \lambda\right)\right\}$ is recursively enumerable;

(iii) a recursive function $k : \bar{I} \rightarrow \mathbb{N}$ exists such that for every $\lambda \in \bar{I}$

$$\lambda^+ A = W_{k(\lambda)}; \text{ and}$$

(iv) there is a recursive function $k : X \times \mathbb{N} \rightarrow \bar{I}$ such that for every $x \in X$

$$A(x) = \bigvee \{k(x, n) \mid n \in \mathbb{N}\}.$$

Proof (i)⇒(ii) Suppose that there is a recursive function $h : X \times \mathbb{N} \rightarrow \bar{I}$ that is increasing with respect to its second variable such that $A(x) = \lim_{n \to +\infty} h(a, n)$. Let us define a function $g : X \times \bar{I} \rightarrow \mathbb{N}$ as follows:

$$g(x, \lambda) = \begin{cases} 1, & \text{when } h(x, n) > \lambda \text{ for some } n \in \mathbb{N}, \\ \perp, & \text{otherwise.} \end{cases}$$

Also assume that $(x, \lambda) \in K(A)$. Then, this is equivalent to $A(x) > \lambda$, which implies that $\lim_{n \to +\infty} h(x, n) > \lambda$; this, in turn, is equivalent to saying that there is an $n \in \mathbb{N}$ such that

$h(x, n) > \lambda$. This last statement is equivalent to the statement that (x, λ) is a member of the domain of g. In different words, $K(A)$ is the domain of a recursive function and because of Theorem 2.5.1 (see p. 36), it is recursively enumerable.

(ii)⇒(iii) Suppose that g is a partial recursive function whose domain is $K(A)$. Because of the s-m-n theorem (i.e., Theorem 2.4.1), there is a recursive function k such that $\varphi_{k(\lambda)}(x) = g(x, \lambda)$. Now $x \in {}^{\lambda+}A$ if and only if $(x, \lambda) \in K(A)$, which means that g converges in (x, λ). In turn, this means that $\varphi_{k(\lambda)}$ converges in x and, thus, $x \in W_{k(\lambda)}$.

(iii)⇒(iv) First note that

$$A(x) = \bigvee \left\{ \lambda \mid \left(\lambda \in \bar{I} \right) \wedge \left(x \in {}^{\lambda+}A \right) \right\} = \bigvee \left\{ \lambda \mid \left(\lambda \in \bar{I} \right) \wedge \left(x \in W_{k(\lambda)} \right) \right\}.$$

Now, set

$$W(\lambda, j) = \left\{ x \mid \left(x \in X \right) \wedge \left(\varphi_{k(\lambda)} \text{ converges in } x \text{ in less than } j \text{ steps} \right) \right\}$$

and let $\pi : \mathbb{N} \to \mathbb{N} \times \bar{I}$ be a recursive one–one function and

$$g(x, n) = \begin{cases} \lambda, & \text{when } \pi(n) = (j, \lambda) \text{ and } x \in W(\lambda, j), \\ 0, & \text{otherwise.} \end{cases}$$

Then, $g(x, n)$ is recursive and $A(x) = \bigvee \{ g(x, n) \mid n \in \mathbb{N} \}$.

(iv)⇒(i) Suppose that k is a recursive function such that

$$A(x) = \bigvee \left\{ k(x, n) \mid n \in \mathbb{N} \right\}$$

and let $h(x, n) = k(x, 1) \vee \cdots \vee k(x, n)$. Then, h is a recursive map increasing with respect to n and $A(x) = \lim_{n \to +\infty} h(a, n)$. □

A recursively enumerable set is not necessarily an α-cut of a fuzzily recursively enumerable fuzzy set as the following result shows:

Theorem 4.6.1 *Assume that $A \subseteq X$. Then, A is an α-cut of a fuzzily recursively enumerable fuzzy set if and only if $A \in \Pi_2^0$.*

Proof Suppose that S is a fuzzily recursively enumerable fuzzy subset and that $S(x) = \lim_{n \to +\infty} h(a, n)$, where h is a recursive function increasing with respect to its second parameter. Then for any $\lambda \in \bar{I}$: $x \in {}^{\lambda+}S$ if and only if $S(X) \geq \lambda$. This can be expressed as $\lim_{n \to +\infty} h(a, n) \geq \lambda$. This simply means that for all k there is an m such that $h(x, m) \geq \lambda - 1/k$ or in mathematical notation

$$(\forall k)(\exists m)\left(h(x, m) \geq \lambda - 1/k \right),$$

which is obviously a Π_2^0 statement. This means that the α-cut of a fuzzily recursively enumerable fuzzy set is a Π_2^0 set.

Assume that $X \supseteq A \in \Pi_2^0$. Then, it is possible to give an example of a fuzzily recursively enumerable fuzzy set S that has a strong α-cut ${}^{\lambda+}S$ that is Π_n^0-complete (see Definition 2.5.8). Now A is one–one reducible to ${}^{\lambda+}S$ by a recursive function $f : X \to X$, that is, $x \in A \Leftrightarrow S(f(x)) \geq \mu$. This means that A is the α-cut of the fuzzily recursively enumerable fuzzy set $S \circ f$. □

Since the notion of fuzzily recursively enumerable sets has been introduced, it makes sense to use it to introduce the counterparts of the notions that are related to recursive enumerability:

Definition 4.6.3 A fuzzy subset A of X is called *fuzzily co-recursively enumerable*[7] if its complement, $\bar{A}$, is fuzzily recursively enumerable. In addition, A is *fuzzily decidable* if it is both fuzzily recursively enumerable and fuzzily co-recursively enumerable.

Proposition 4.6.3 *A fuzzy subset* $A : X \rightarrow [0,1]$ *is fuzzily co-recursively enumerable if and only if there is a recursive function* $k : X \times \mathbb{N} \rightarrow \bar{I}$ *such that for all* $x \in X$, $k(x,n)$ *is decreasing with respect to* n *and*

$$A(x) = \lim_{n \to +\infty} k(x,n).$$

Proof Assume that $l : X \times \mathbb{N} \rightarrow \bar{I}$ is a recursive function that is increasing with respect to its second argument and $\bar{A}(x) = \lim_{n \to +\infty} l(x,n)$. Then,

$$A(x) = 1 - \lim_{n \to +\infty} l(x,n) = \lim_{n \to +\infty} \left(1 - l(x,n)\right) = \lim_{n \to +\infty} k(x,n),$$

where $k(x,n) = 1 - l(x,n)$. Clearly, k is recursive and decreasing. Similarly, one can prove the converse. $\qquad\square$

With the following result it is possible to give a characterization of fuzzily decidable subsets.

Theorem 4.6.2 *A fuzzy subset* $A : X \rightarrow [0,1]$ *is fuzzily decidable if and only if for every* $x \in X$, $A(x)$ *is the limit of a classically computable nested sequence of intervals, that is, if and only if there are two recursive functions* $k : X \times \mathbb{N} \rightarrow \bar{I}$ *and* $l : X \times \mathbb{N} \rightarrow \bar{I}$ *such that for any* $x \in X$:

- k *is increasing and* l *is decreasing with respect to their second argument;*

- *for every* $n \in \mathbb{N}$, $k(x,n) \leq A(x) \leq l(x,n)$; *and*

- $\lim_{n \to +\infty} k(x,n) = A(x) = \lim_{n \to +\infty} l(x,n)$.

Proof The first part can be proved by noticing that A is fuzzily decidable which means that it is both fuzzily recursively enumerable and co-recursively enumerable. Remark 4.6.1 leads us to the conclusion that $k(x,n) \leq A(x)$. Since

$$A(x) = \bigwedge \{l(x,n) \mid n \in \mathbb{N}\},$$

because l is decreasing with respect to n, one concludes that $A(x) \leq l(x,n)$. The third part is obvious. $\qquad\square$

---

7. Gerla [49] uses instead the term *recursively co-enumerable*, which does not make any sense to this author.

A recursive function $f : X \times \mathbb{N} \to \bar{I}$ is *recursive convergent to* a fuzzy subset $A : X \to [0,1]$ when for any $x \in X$, $A(x) = \lim_{n \to +\infty} f(x,n)$ and there is a recursive function $g : X \times \mathbb{N} \to \mathbb{N}$ such that for all $x \in X$ and $p \in \mathbb{N}$

$$|f(x,n) - f(x,n)| < \frac{1}{p}, \text{for any } n, m \geq g(x,p).$$

Equipped with the notion just described, it is possible to have an alternative characterization of fuzzy decidability:

Theorem 4.6.3 *A fuzzy subset* $A : X \to [0,1]$ *is fuzzily decidable if and only if there is a recursive function* $f : X \times \mathbb{N} \to \bar{I}$ *which is recursive convergent to A.*

Proof Suppose that A is fuzzily decidable and that k and l are as in Theorem 4.6.2. In essence, the theorem claims that for all $n, m \in \mathbb{N}$ and $x \in X$

$$|k(x,n) - k(x,m)| \leq l(x, n \wedge m) - k(x, n \wedge m).$$

Indeed, it holds that

$$l(x, n \wedge m) \geq l(x,n) \geq k(x,n) \text{ and } -k(x,m) \leq -k(x, n \wedge m),$$

and so for $n \geq m$

$$\begin{aligned} |k(x,n) - k(x,m)| &= k(x,n) - k(x,m) \\ &\leq l(x, n \wedge m) - k(x,m) \\ &\leq l(x, n \wedge m) - k(x, n \wedge m). \end{aligned}$$

When $m \geq n$, it easy to see that similar things hold true. Let

$$g(x,p) = \min\left\{ j \in \mathbb{N} \,\middle|\, l(x,j) - k(x,j) < \frac{1}{p} \right\}.$$

This function is clearly recursive and

$$|k(x,n) - h(x,m)| \leq l(x, nn \wedge m) - k(x, n \wedge m) < \frac{1}{p} \text{ for all } m, n \geq g(x,p)$$

which proves that the recursive function k is recursive convergent to A.

Now let us assume that f is a recursive function that is recursive convergent to A by the recursive function g. Also, define, for all $n \in \mathbb{N}$, $m_n = g(x,n)$. It holds that

$$f(x, m_n) - \frac{1}{n} \leq A(x) \leq f(x, m_n) + \frac{1}{n}.$$

If we set

$$u(x,n) = \bigvee\left\{ f(x, m_i) - \frac{1}{i} \,\middle|\, i = 1, ..., n \right\} \text{ and } v(x,n) = \bigwedge\left\{ f(x, m_i) - \frac{1}{i} \,\middle|\, i = 1, ..., n \right\},$$

then u is recursive and increasing with respect to n while v is also recursive but decreasing with respect to n. In addition, because

$$\lim_{n \to +\infty} \left(f(x, m_n) - \frac{1}{n} \right) = A(x) = \lim_{n \to +\infty} \left(f(x, m_n) + \frac{1}{n} \right),$$

it holds that

$$\lim_{n \to +\infty} u(x, n) = A(x) = \lim_{n \to +\infty} v(x, n).$$

$\square$

An Extension of the Arithmetic Hierarchy   Biacino and Gerla [15] extended the arithmetic hierarchy by allowing fuzzy relations to be part of different classes of relations. In particular, they defined the various classes as follows:

- $\Sigma_0^0 = \Pi_0^0$, where these classes are equal to the class of fuzzy relations that are characterized by computable functions whose codomain is the set $\bar{I}$;

- the class Σ_{n+1}^0 contains all projections of fuzzy relations in Π_n^0;

- the class Π_{n+1}^0 contains all co-projections of fuzzy relations in Σ_n^0; and

- $\Delta_n^0 = \Sigma_n^0 \cap \Pi_n^0$.

In order to fully understand what is involved, one must realize that in many-valued logics, the existential and universal quantifiers are "replaced" by the projection and co-projection "operators." The following result shows how the basic results of the arithmetic hierarchy are *extended* in a fuzzy setting:

Proposition 4.6.4 *For every n that is greater or equal to one:*

(i) the projection (co-projection) of an element of Σ_n^0 (Π_n^0) belongs to Σ_n^0 (Π_n^0);

(ii) $\Sigma_n^0 \cup \Pi_n^0 \subseteq \Delta_{n+1}^0$ but $\Sigma_n^0 \cup \Pi_n^0 \neq \Delta_n^0$;

(iii) Σ_n^0 and Π_n^0 are both closed under finite unions and intersections;

(iv) $\Sigma_n^0 = \left\{ \bar{R} \mid R \in \Pi_n^0 \right\}$ and $\Pi_n^0 = \left\{ \bar{R} \mid R \in \Sigma_n^0 \right\}$; and

(v) Δ_n^0 is closed under finite unions and intersections and the complement operation.

The following definition is needed for the result that follows:

Definition 4.6.4 Assume that $f : \mathbb{N}^m \to [0,1]$ is a function and $n \leq m$. Then, function $\lim^n f : \mathbb{N}^{m-n} \to [0,1]$ is defined as follows:

$$\lim{}^1 f(x_1, ..., x_{m-1}) = \lim_{x \to +\infty} f(x_1, ..., x_{m-1}, x)$$

$$\lim{}^n f(x_1, ..., x_{m-n}) = \lim{}^1 \left(\lim{}^{n-1} f(x_1, ..., x_m) \right).$$

Theorem 4.6.4 *For all $n \geq 1$ a fuzzy relation R in $\mathbb{N}$ belongs to Σ_n^0 or Π_n^0 if and only if, for all $x \in \mathbb{N}^m$,*

$$R(x) = \lim{}^n h(x, y_1, ..., y_n), \tag{4.8}$$

where $h : \mathbb{N}^m \times \mathbb{N}^n \to \bar{I}$ is computable and $h(x, y_1, ..., y_n)$ is increasing or decreasing, respectively, with respect to y_i when i is odd and decreasing or increasing, respectively, otherwise.

Proof The proof is by induction on n. Suppose that $R \in \Sigma_1^0$. Then, R is the projection of an element $R' \in \Pi_0^0$, that is, $R(x) = \bigvee \{R'(x, y) \mid y \in \mathbb{N}\}$. This implies that (4.8) holds with $h(x, y) = \bigvee \{R'(x, y') \mid y' \leq y\}$. Similarly, if $R \in \Pi_1^0$, then R is the co-projection of an element $R' \in \Sigma_0^0$, that is, $R(x) = \bigwedge \{R'(x, y) \mid y \in \mathbb{N}\}$ and so R is the limit of the sequence $k(x, y) = \bigwedge \{R'(x, y') \mid y' \leq y\}$. Assume that $R \in \Sigma_1^n$, where $n > 1$. Then, R is the projection of an element $R' \in \Pi_{n-1}^0$, that is, $R(x) = \bigvee \{R'(x, y) \mid y \in \mathbb{N}\}$. By the induction hypothesis, there is a computable function $h' : \mathbb{N}^m \times \mathbb{N} \times \mathbb{N}^{n-1} \to \bar{I}$ such that $R'(x, y_1) = \lim{}^{n-1} h'(x, y_1, y_2, ..., y_n)$. In addition, for all $i \geq 2$, function $h'(x, y_1, y_2, ..., y_n)$ is decreasing with respect to y_i if i is even, increasing otherwise. Let

$$h(x, y_1, y_2, ..., y_n) = \bigvee \left\{ h'(x, y, y_2, ..., y_n) \,\middle|\, y \leq y_1 \right\}.$$

Then $h(x, y_1, y_2, ..., y_n)$ is computable, increasing with respect to y_i when i is odd, decreasing when i is even. We have

$$\lim{}^n h(x, y_1, y_2, ..., y_n) = \lim{}^1 \left(\lim{}^{n-1} h(x, y_1, y_2, ..., y_n) \right)$$

$$= \lim{}^1 \left(\lim{}^{n-1} \left(\bigvee \left\{ h'(x, y, y_2, ..., y_n) \,\middle|\, y \leq y_1 \right\} \right) \right)$$

$$= \lim{}^1 \left(\bigvee \left\{ \lim{}^{n-1} h'(x, y, y_2, ..., y_n) \,\middle|\, y \leq y_1 \right\} \right)$$

$$= \lim{}^1 \left(\bigvee \left\{ R'(x, y) \,\middle|\, y \leq y_1 \right\} \right)$$

$$= \bigvee \left\{ R'(x, y) \,\middle|\, y_1 \in \mathbb{N} \right\}$$

$$= r(x).$$

This sequence of equalities makes use of the fact that if $(\nu_n)_{n \in \mathbb{N}}$ and $(\mu_n)_{n \in \mathbb{N}}$ are two sequences of elements of $[0, 1]$ both decreasing or increasing, then

$$\lim_{n \to +\infty} (\nu_n \vee \mu_n) = (\lim_{n \to +\infty} \nu_n) \vee (\lim_{n \to +\infty} \mu_n),$$

$$\lim_{n \to +\infty} (\nu_n \wedge \mu_n) = (\lim_{n \to +\infty} \nu_n) \wedge (\lim_{n \to +\infty} \mu_n).$$

In order to prove the case $R \in \Pi_n^0$, one has to work in a similar way. The converse of this result is obvious. □

Class Δ_2^0 is particularly interesting since it includes the so-called *limiting* recursive predicates, sets, etc. (see [126, Chap. 3] for details). In our particular case, a fuzzy subset A :

$X \rightarrow [0, 1]$ is *limit decidable* if there is a recursive function $h : X \times \mathbb{N} \rightarrow \bar{\mathbb{I}}$ such that $A(x) = \lim_{n\rightarrow+\infty} h(x, n)$ for all $x \in X$. The set A is *by-frequency decidable* when there is a recursive function $k : X \times \mathbb{N} \rightarrow \{0, 1\}$ such that for all $x \in X$

$$A(x) = \lim_{n\rightarrow\infty} \left(k(x, 1) + \cdots + k(x, n) \right) \Big/ n.$$

Also, the fuzzy subset A is *with limit decidable cuts* when there is a computable function $c : X \times \bar{\mathbb{I}} \times \mathbb{N} \rightarrow \{0, 1\}$ such that

$$A(x) > \lambda \text{ implies that } \lim_{n\rightarrow+\infty} c(x, \lambda, n) = 1 \text{ and}$$

$$A(x) < \lambda \text{ implies that } \lim_{n\rightarrow+\infty} c(x, \lambda, n) = 0.$$

(4.9)

The following result gives a partial characterization of the class Δ_2^0.

Proposition 4.6.5 *The following are equivalent:*

(i) $A \in \Delta_2^0$;

(ii) A is limit decidable;

(iii) A is by-frequency decidable; and

(iv) A is with limit decidable cuts.

Proof (i)$\Rightarrow$(ii) If $A \in \Delta_2^0$, then because of Theorem 4.6.4, the following holds true:

$$A(x) = \lim_{n\rightarrow+\infty} \left(\lim_{m\rightarrow+\infty} h(x, n, m) \right) = \lim_{n\rightarrow+\infty} \left(\lim_{m\rightarrow+\infty} h'(x, n, m) \right),$$

where $h(x, n, m)$ and $h'(x, n, m)$ are computable, increasing or decreasing, respectively, with respect to n and decreasing or increasing, respectively, with respect to m. Let

$$\ell_n(x) = \lim_{m\rightarrow+\infty} h(x, n, m) \text{ and } \ell'_n(x) = \lim_{m\rightarrow+\infty} h'(x, n, m).$$

Then, for every fixed $x \in \mathbb{N}$,

$$(\forall \varepsilon > 0)\, (\exists v_\varepsilon)\, (\forall n \geq v_\varepsilon)\, \left| \ell_n(x) - \ell'_n(x) \right| < \frac{\varepsilon}{3}.$$

In addition,

$$(\forall n \in \mathbb{N})\, (\forall \varepsilon > 0)\, (\exists m_{\varepsilon,n})\, (\forall m \geq m_{\varepsilon,n}) \left(\left| \ell_n(x) - h(x, n, m) \right| < \varepsilon/3 \right.$$

$$\left. \text{and } \left| \ell'_n(x) - h'(x, n, m) \right| < \varepsilon/3 \right).$$

Thus,

$$(\forall \varepsilon > 0)\, (\forall n \geq v_\varepsilon)\, (\forall m \geq m_{\varepsilon,n})\, \left| h(x, n, m) - h'(x, n, m) \right| \leq$$

$$\left| h(x, n, m) - \ell_n(x) \right| + \left| \ell_n(x) - \ell'_n(x) \right| + \left| \ell'_n(x) - h'(x, n, m) \right| < \varepsilon.$$

With this, one can construct two strictly increasing sequences $\left(n_k(x)\right)$ and $\left(m_k(x)\right)$ that have the following property:

$$\left|h\left(x, n_k(x), m_k(x)\right) - h'\left(x, n_k(x), m_k(x)\right)\right| < 1/k \text{ for all } k \in \mathbb{N}. \tag{4.10}$$

Let

$$u(x, k) = h\left(x, n_k(x), m_k(x)\right) \text{ and } u'(x, k) = h'\left(x, n_k(x), m_k(x)\right).$$

Then, one has to prove that $A(x) = \lim_{k \to +\infty} u(x, k) = \lim_{k \to +\infty} u'(x, k)$. Indeed, since

$$u(x, k) = h\left(x, n_k(x), m_k(x)\right) \geq \bigwedge_m h\left(x, n_k(x), m\right)$$

and

$$u'(x, k) = h'\left(x, n_k(x), m_k(x)\right) \leq \bigvee_m h'\left(x, n_k(x), m\right),$$

it follows that

$$\lim{}' u(x, k) \geq \lim{}'\left(\bigwedge_m h\left(x, n_k(x), m\right)\right) = A(x)$$

$$= \lim{}''\left(\bigvee_m h'\left(x, n_k(x), m\right)\right)$$

$$\geq \lim{}'' u'(x, k),$$

where

$$\lim{}' \alpha_n = \bigvee_m \bigwedge_{n \geq m} \alpha_n \text{ and } \lim{}'' \alpha_n = \bigwedge_m \bigvee_{n \geq m} \alpha_n.$$

From these inequalities and (4.10) it follows that $\lim'' u'(x, k) = A(x) = \lim' u(x, k)$. Since (4.10) implies that $\lim'' u'(x, k) = \lim'' u(x, k)$ and $\lim' u'(x, k) = \lim' u(x, k)$, it follows that A is limit decidable.

(ii)$\Rightarrow$(iii) Suppose that $A(x) = \lim_{n \to +\infty} h(x, n)$, where h is computable. Then, it follows that $A(x) = \lim_{n \to +\infty} \frac{1}{n} \sum_{i \leq n} h(x, i)$. Let $\lfloor x \rfloor$ be the integer part of the real number x and define the function $k : \mathbb{N} \times \mathbb{N} \to \{0, 1\}$ as follows:

$$k(x, 1) = \lfloor h(x, 1) \rfloor,$$

$$k(x, n + 1) = \left\lfloor \sum_{i \leq n+1} h(x, i) \right\rfloor - \sum_{i \leq n} k(x, i).$$

Clearly, $\sum_{i \leq n} k(x, i) = \lfloor \sum_{i \leq n} h(x, i) \rfloor$. In addition, $k(x, n) \in \{0, 1\}$ for all $x, n \in \mathbb{N}$. Indeed

$$k(x, n + 1) = \left\lfloor \sum_{i \leq n+1} h(x, i) \right\rfloor - \sum_{i \leq n} k(x, i)$$

$$= \left\lfloor \sum_{i \leq n} h(x, i) + h(x, n + 1) \right\rfloor - \left\lfloor \sum_{i \leq n} h(x, i) \right\rfloor \in \{0, 1\}.$$

Since

$$A(x) = \lim_{n \to +\infty} \frac{1}{n} \sum_{i \leq n} h(x, i) = \lim_{n \to +\infty} \frac{1}{n} \left[\sum_{i \leq n} h(x, i) \right] = \lim_{n \to +\infty} \frac{1}{n} \sum_{i \leq n} k(x, i),$$

clause (iii) is proved.

(iii)$\Rightarrow$(iv) Let $h(x, n) = \frac{1}{n} \sum_{i \leq n} k(x, i)$. Then, $A(x) = \lim_{n \to +\infty} h(x, n)$ and so, for all $\alpha \in \bar{I}$

$$\left(A(x) < \alpha \right) \Rightarrow (\exists v)(\forall n \geq v)\left(h(x, n) < \alpha \right)$$

and

$$\left(A(x) > \alpha \right) \Rightarrow (\exists v)(\forall n \geq v)\left(h(x, n) > \alpha \right).$$

Now, set

$$c(x, \alpha, n) = \begin{cases} 1, & \text{when } h(x, n) \geq \alpha, \\ 0, & \text{otherwise.} \end{cases}$$

Then c is computable and (4.9) holds true.

(iv)$\Rightarrow$(i) Suppose that (4.9) holds true. Then $A(x) \geq \alpha$ because $\lim_{m \to +\infty} c(x, \alpha, m) = 1$. In addition,

$$A(x) = \bigvee \left\{ \alpha \in \bar{I} \,\Big|\, \lim_{m \to +\infty} c(x, \alpha, m) = 1 \right\} = \bigvee_{n \in \mathbb{N}} \left\{ \alpha_n \,\Big|\, (\exists v)(\forall m \geq v)c(x, \alpha_n, m) = 1 \right\},$$

where (α_n) is a coding of $\bar{I}$. Set

$$r'(x, n, m) = \begin{cases} \alpha_n, & \text{when } c(x, \alpha_n, m) = 1 \\ 0, & \text{otherwise} \end{cases} \qquad r(x, n, m, v) = \begin{cases} 1, & \text{when } m < v, \\ r'(x, n, m), & \text{otherwise.} \end{cases}$$

Then,

$$A(x) = \bigvee_{n \in \mathbb{N}} \bigvee_{v \in \mathbb{N}} \bigwedge_{m \geq v} r'(x, n, m) = \bigvee_{n \in \mathbb{N}} \bigvee_{v \in \mathbb{N}} \bigwedge_{m \in \mathbb{N}} r(x, n, m, v).$$

Since the projection of an element of Σ_2^0 belongs to Σ_2^0, it follows that $A \in \Sigma_2^0$. Similarly, one can prove that $A \in \Pi_2^0$. $\qquad \square$

4.7   W-Recursivity

In Sect. 4.3.5 I presented an extension of Turing W-machines that forms the basis for a theory of fuzzily recursive functions, which are dubbed W-recursive functions. In essence, the ideas presented in this section are the fuzzy theoretic equivalent of general recursive functions (see Sect. 2.4).

Definition 4.7.1 Assume that $g_i : (U^*)^{k_i} \times V^* \to W, i = 1, 2, ..., n$ and $f : (V^*)^n \times X^* \to W$ are W-valued functions. Then, the operation that associates to f and $g = (g_1, g_2, ..., g_n)$ the function

$$h : [(U^*)^{k_1} \times (U^*)^{k_2} \times \cdots \times (U^*)^{k_n} \times X^* \to W$$

defined by

$$h(x \mid u^k) = \bigoplus_{v^n \in (V^*)^n}^* \left\{ f(x \mid v^n) \otimes \left[\bigotimes_{i=1}^n {}^* g_i(v_i \mid u_i^{k_i}) \right] \right\},$$

where $\otimes^*$ is the extension of the operator $\otimes$ to the set of calculi with input (u, q) and output (q', v), $k = \sum_{i=1}^n k_i, v \in V^*, x \in X^*$, and $u^k = (u_1^{k_1}, ..., u_n^{k_n}) \in (U^*)^{k_1} \times \cdots \times (U^*)^{k_n}$, is called W-composition and is denoted as $\circ$. Thus, $h = f \circ g$.

The operation of W-minimalization is defined in terms of the codifier and decodifier functions, which were introduced in Definition 2.2.14 on p. 24.

Definition 4.7.2 Assume that $g : [(U^*)^k \times V^*] \times X^* \to W$, where $k \geq 1$, is a W-valued function. Then, the operation that associates to g the function $f : (U^*)^k \times V^* \to W$, which is defined as follows:

$$f(v \mid u^k) = g(x_0 \mid u^k, v)$$

provided that

$$v = \mathscr{W} \left(\bigvee_{i \geq q} \{ i \in \mathbb{N} \mid g(x_0 \mid u^k, \mathscr{W}(i)) = g(x_0 \mid u^k, v) \} \right),$$

where $x_0 \in X^*$ and $q \in \mathbb{N}$, is called W-minimalization and is written as $f = \min_{x_0, q} g$.

With W-minimalization it is possible to define partially W-computable functions from totally W-computable functions. Nevertheless, it is possible to get a total function. In this case, function g is called W-regular.

Definition 4.7.3 Assume that $f : (U^*)^k \times U^* \to W$ and $g : (U^*)^{k+2} \times U^* \to W$ are W-valued functions. Then, the operation that associates to them the W-valued function $h : (U^*)^{k+1} \times U^* \to W$ defined by

$$h(v \mid u^k, \mathscr{W}(q)) = f(v \mid u^k),$$

$$h(v \mid u^k, \mathscr{W}(\mathscr{N}(x) + 1)) = \bigoplus_{u \in U^*}^* \{ g(v \mid u^k, y, x) \otimes h(y \mid u^k, x) \},$$

where $q \in \mathbb{N}, x \in U^*, \mathscr{N}(x) \geq q$, is called primitive W-recursion.

Clares [30] has proved that the set of W-computable functions is closed under the operations of W-composition and W-primitive recursion. In addition, the same author proved that the set of practically W-computable functions is closed under the operation of W-minimalization. Also, one needs to define the successor, the zero, and the projection functions in order to be able to define primitive and general W-recursive functions. Obviously, the zero and the successor functions cannot be fuzzified and are redefined as follows:

$$\mathscr{Z}(v \mid u) = \begin{cases} 1, & \text{if } v = 0, \\ 0, & \text{otherwise,} \end{cases}$$

$$\mathscr{S}(v \mid u) = \begin{cases} 1, & \text{if } v = u + 1, \\ 0, & \text{otherwise,} \end{cases}$$

for all $v, u \in \mathbb{N}$. The projection function is redefined in terms of a computable function $\psi : \mathbb{N} \times \mathbb{N} \to W$, which is an element of the set

$$\Psi = \left\{ \varphi \mid \varphi : \mathbb{N} \times \mathbb{N} \to W \text{ and } \varphi \text{ is computable} \right\}.$$

This set contains functions that are generalizations of Kronecker's δ function:

$$\delta(v \mid u) = \begin{cases} 1, & \text{if } v = u, \\ 0, & \text{otherwise.} \end{cases}$$

Clearly, $\delta \in \Psi$ since δ is computable. Thus, the projection function is redefined as follows:

$$\mathscr{P}_{\psi}^{n,i}(v \mid u_1, u_2, ..., u_n) = \psi(v \mid u_i), v, u_1, ..., u_n \in \mathbb{N}, \psi \in \Psi.$$

Since there are many different functions that make up the set Ψ, it is necessary to specify which function is used in a particular realization of the projection function.

The following lemmata, which will be used later on in this section, have been proved in [30][8]:

Lemma 4.7.1 *Assume that* $N = \{i_1, ..., i_n\}$, $M = \{j_1, ..., j_m\}$, *and* $K = \{k_1, ..., k_n\}$ *are sets of positive integer numbers, where* $n, m \geq 1$. *If* $M \cap N = \varnothing$, *there is a multitape Turing machine* $C_K^{N,M}$, *with at least* $n + m$ *tapes, such that when it starts with the expressions*

$$\overline{u}_{i_r} \in (C^*)^{k_r'}, k_r' \geq k_r,$$

where

$$\overline{u}_{i_r} = (u_{1 i_r}, u_{2 i_r}, ..., u_{k_r' i_r}), r = 1, 2, ..., n,$$

on the tapes designated by the set N *and all remaining expression on the remaining tapes, it stops with the expression*

$$\overline{v} \in (C^*)^{\sum_{r=1}^{n} k_r},$$

where

$$\overline{v} = (u_{1 i_1}, u_{2 i_2}, ..., u_{k_i i_1}, u_{k_2 i_2}, ..., u_{1 i_n}, u_{2 i_n}, ..., u_{k_n i_n})$$

on all tapes designated by the set M.

Clares dubbed $C_K^{N,M}$ a "generalized copying machine." When the expressions on the tapes designated by N are copied on the tapes designated by M, the machine is referred as $C^{N,M}$, that is, the K is dropped from the name.

---

8. The proofs are omitted since the original text is in Spanish and my knowledge of this language is nonexistent.

Lemma 4.7.2 *Let $Z_1 = (Q_1, U, V_1, S, \delta_1, \pi_1, \eta^{T_1})$ and $Z_2 = (Q_2, V_1, V, S, \delta_2, \pi_2, \eta^{T_2})$ be two W-Turing machines. Then, if $Q_1 \cap Q_2 = \varnothing$ and $(U \cup V_1 \cup V) \subset C$, there is a W-Turing machine $Z = (Q, U, V, S, \delta, \pi, \eta^T)$ such that for all $u \in U$ and for all $v \in V$*

$$p(v \mid u) = \bigoplus_{v' \in V_i^*}^{*} \left\{ p_1(v' \mid u) \otimes \bigoplus_{a_0 \vdash^* a_n}^{*} \left[\pi_s \otimes \bigotimes_{i=0}^{n-1}{}^{*} p^{Z_2}(a_{i+1} \mid a_i) \right] \right\},$$

where $a_i \in D(Z_2)$, $a_0 = \xi q \gamma$, $a_n = \xi q' \rho$, for all $n \in \mathbb{N}$, $q, q' \in S_2$, $s \in T_2$, $v' = \xi \gamma$, and $v = \xi \rho$.

In the rest of this section the notation $Z_1 \to Z_2$ will denote machine Z, that is, machine Z_1 followed by Z_2.

Lemma 4.7.3 *Let $Z_1^m = (Q_1, U, V_1, S, \delta_1, \pi_1, \eta^{T_1})$ and $Z_2^n = (Q_2, V_1, V, S, \delta_2, \pi_2, \eta^{T_2})$ be two W-Turing machines with m and n tapes, respectively, such that $m \geq n \geq 1$ and let J be the following set*

$$J = \left\{ i_j \mid i \leq i_j \leq m, \; i_j \neq i_k, \; j, k = 1, 2, ..., n \right\}.$$

Then, if $S_1 \cap S_2 \neq \varnothing$ and $V_1 \subseteq V$, there is a W-Turing machine $Z^m = (Q, U, V, S, \delta, \pi, \eta^T)$ with m tapes such that

$$p(v \mid u) = \bigoplus_{t' \in (V'^*)^n}^{*} \left\{ p_2(t \mid v') \otimes p_1(v' \mid u) \right\},$$

where $t'_j = v'_{i_j}$, $t_j = v_{i_j}$ for $j = 1, 2, ..., n$, and $v_i = v'_i$ for $i \notin J$, $1 \leq i \leq m$.

As in the previous case, for the rest of this section, the notation $Z_1 \to_J Z_2$ will denote machine Z^m.

Equipped with these results, it is possible to prove a few important theorems.

Theorem 4.7.1 *Assume that $n, i \in \mathbb{N}$, $i \geq n$, and $\psi \in \Psi$. Then, $\mathscr{P}_{\psi}^{n,i}$ is a W-computable function.*

Proof Suppose that $Z_{\psi}^{(2)}$ is W-Turing machine with two tapes whose input and output are printed on the first tape while the second one is exclusively used to compute function ψ. Then, the following machine

$$Z = \mathscr{D}^{(1)} \to \mathscr{D}^{(0)} \xrightarrow{i-1} \mathscr{D}^{(1)} \to C_1^{1,2} \xrightarrow{\bullet} Z_{\psi}^{(2)} \xrightarrow{\bullet} C^{2,1}$$

W-computes function $\mathscr{P}_{\psi}^{n,i}$, where $\mathscr{D}^{(1)}$ is the deterministic Turing machine that moves one word to the right on tape 1.   □

Definition 4.7.4 A function $f : \mathbb{N}^n \times \mathbb{N} \to W$ is *primitive W-recursive* if it can be formed from the functions $\mathscr{Z}$, $\mathscr{S}$, and $\mathscr{P}_{\psi}^{n,i}$, $\psi \in \Psi$, by applying the operations of W-composition and primitive W-recursion a finite number of times.

Clearly, any function $f : \mathbb{N}^n \times \mathbb{N} \to W$ that is primitive W-recursive is a total W-computable function. In addition, one sees that the set of primitive W-recursive functions is closed under the operations of W-composition and primitive W-recursion.

Definition 4.7.5 A function $f : \mathbb{N}^n \times \mathbb{N} \to W$ is a partially W-recursive function when it can be formed from the functions $\mathscr{Z}$, $\mathscr{S}$, and $\mathscr{P}_{\psi}^{n,i}$, $\psi \in \Psi$, by applying the operations of W-composition, primitive W-recursion, and W-minimalization a finite number of times and ψ may be a different function each time $\mathscr{P}_{\psi}^{n,i}$ is used.

If W-minimalization is applied to W-regular functions, then f is a total function and is called W-recursive. In addition, from the previous definition, it should be clear that the set of partially W-recursive functions is closed under the operations of W-compositions, primitive W-recursion, and W-minimalization. Furthermore, it is almost trivial to show that when $f : \mathbb{N}^n \times \mathbb{N} \to W$ is a (partially) W-recursive function, then it is a (partially) W-computable function.

Any computing device has its limits; nevertheless, the Church–Turing thesis is the claim that there is a limit to what can be accomplished by any computing device. In particular, this limit is dictated by what can be achieved by the Turing machine. Thus, if the validity of this thesis is taken for granted, then no fuzzy computing device should be able to surpass the computational power of the Turing machine, but it was shown that in general this is not the case (see Sect. 4.4). Regardless of this result, which was shown almost 15 years later, Clares and Delgado, who wanted to extend classical recursion theory, had to define the analogue of the Church–Turing thesis. Indeed, they had formulated a *soft* Church–Turing thesis that can be stated as follows: *Every W-computable function is W-recursive and vice versa.* The following notion of equivalence between W-functions is the way to formally express the intuitive notion of equivalence just described:

Definition 4.7.6 Assume that $f, g : \mathbb{N}^n \times \mathbb{N} \to W$ are two W-computable functions that have a common domain $D \subseteq \mathbb{N}^n$. Then, f is *equivalent* to g if

$$f(v \mid u_1, u_2, ..., u_n) = g(v \mid u_1, u_2, ..., u_n),$$

for all $(u_1, u_2, ..., u_n) \in D$ and for all $v \in \mathbb{N}$.

The proof of the result that follows is based on the assumed validity of the soft thesis, which, in turn, is based on the assumption that the crisp thesis is valid.

Theorem 4.7.2 *For every W-computable function $f : \mathbb{N}^n \times \mathbb{N} \to W$ there is another equivalent function $g : \mathbb{N}^n \times \mathbb{N} \to W$ that is W-recursive.*

Proof Assume that f' is a function defined as follows:

$$f'(v \mid u_1, u_2, ..., u_n) = \begin{cases} 1, & \text{when } f(v \mid u_1, u_2, ..., u_n) > 0, \\ 0, & \text{when } f(v \mid u_1, u_2, ..., u_n) = 0. \end{cases}$$

Since f is W-computable, f' is a computable crisp function. Then, because of the soft Church–Turing thesis, f' is recursive and therefore W-recursive.

Assume that $\sigma : \mathbb{N}^n \to \mathbb{N}$ is a total, injective, and computable function such that σ^{-1} is computable in $\sigma(\mathbb{N}^n)$. These two functions are W-recursive in $\mathbb{N}$ and $\sigma(\mathbb{N}^n)$, respectively. Also, assume that $\delta, \psi : \mathbb{N} \times \mathbb{N} \to W$ are such that

(i) δ is Kronecker's function; and

(ii) ψ satisfies

$$\psi(x \mid z) = \begin{cases} f(v \mid u_1, u_2, ..., u_n), & \text{when } x = \sigma(v, \overbrace{y, ..., y}^{n-1}), x = \sigma(v', \overbrace{y, ..., y}^{n-1}) \\ & \text{for } v' \in \mathbb{N}, y = \sigma(u_1, ..., u_n), \\ 0, & \text{otherwise.} \end{cases}$$

Since f is W-computable and σ^{-1} is injective and computable in $\sigma(\mathbb{N})$, ψ is also W-computable. Therefore, $\delta, \psi \in \Psi$. By W-composition, one gets function $g : \mathbb{N}^n \times \mathbb{N} \to W$ that is defined as follows:

$$g = \mathscr{P}_\delta^{n,1} \circ \sigma^{-1} \circ \mathscr{P}_\psi^{n,1} \circ \left(\overbrace{\sigma, ..., \sigma}^{n}\right) \circ \left(f', \overbrace{\sigma, ..., \sigma}^{n-1}\right).$$

This function is equivalent to f.

The proof is not complete since it is not clear whether function σ exists. In order to address this problem, let us consider that $\sigma : \mathbb{N}^n \to \mathbb{N}$ is a function such that $v = \sigma(u_1, u_2, ..., u_n)$ is obtained by the following operations:

(i) u_i, where $i = 1, 2, ..., n$, is represented in a unary numeric system, that is, u_i is represented by a sequence A_i of $u_i + 1$ zeroes;

(ii) the binary sequence $A = 1A_1 1A_2 1...1A_n$ is formed; and

(iii) v is the decimal representation of the bitstring A.

Then σ satisfies all the needed conditions and the theorem holds. □

Remark 4.7.1 For partially W-computable functions, One can show a similar result by using the operation of W-minimalization.

The next result follows from the definition of W-recursiveness:

Theorem 4.7.3 *The set of W-Turing machines is not enumerable.*

   Proof If every W-recursive function is W-computable, there is a W-Turing machine that W-computes it. Since a W-recursive function is formed by applying the projection operator a finite number of times (see Definition 4.7) and the set Ψ is not enumerable, it follows that the set of W-Turing machines is not enumerable. □

Corollary 4.7.1 *The set of computable problems is a proper subset of the set of W-computable problems.*

   Proof Any computable function is W-computable which implies that the computable problems are part of the W-computable problems. In addition, this is a strict inclusion since the set of W-Turing machines is not enumerable. □

Again, this result is based on the assumption that the Church–Turing thesis is valid.

4.8   Effective Domain Theory and Fuzzy Subsets

Effective domain theory has been applied to fuzzy subsets to give alternative definitions of semidecidability and decidability of fuzzy subsets. The work was initiated by Biacino and Gerla [16] and it was continued by Gerla [50, 51]. In what follows, I will first introduce the relevant mathematical notions and then I will describe how these concepts can be used in fuzzy set theory.

4.8.1   On Effective Lattices

For the rest of this section L will stand for a frame (see Definition 3.1.8).

Definition 4.8.1 An element a of an effective lattice $(L, \leqslant, B)$ is semidecidable if the set $\{n \in \mathbb{N} \mid b_n \ll a\}$ is recursively enumerable.

Note that any $b \in B$ is semidecidable. Also, when $L = [0, 1]$, then x is semidecidable if and only if the set $\{r \in \bar{\mathbb{I}} \mid r < x\}$ is recursively enumerable.

Proposition 4.8.1 *Assume that $(L, \leqslant, B)$ is an effective lattice. Then, the following statements are equivalent:*

(i) x is semidecidable;

(ii) there is a recursive function f such that $(b_{f(n)})_{n \in \mathbb{N}}$ preserves the relation $\ll$ and

$$x = \bigvee_{n \in \mathbb{N}} b_{f(n)}; \tag{4.11}$$

(iii) there is a recursive function f such that $(b_{f(n)})_{n \in \mathbb{N}}$ is order-preserving and satisfies (4.11); and

(iv) there is a recursive function f such that $(b_{f(n)})_{n \in \mathbb{N}}$ satisfies (4.11).

Proof (i)$\Rightarrow$(ii) Suppose that x is semidecidable and that $g : \mathbb{N} \to \mathbb{N}$ is a total recursive function whose codomain is the set $\{n \in \mathbb{N} \mid b_n \ll x\}$. Then $x = \bigvee_{n \in \mathbb{N}} b_{g(n)}$ with $b_{g(m)} \ll x$ for any $n \in \mathbb{N}$. Also, there is a recursive function f such that $\mathrm{cod}(f) \subseteq \mathrm{cod}(g)$, where cod returns the codomain of a function, $b_{f(n)} \ll b_{f(n+1)}$, and $b_{g(n)} \ll b_{f(n+1)}$. Indeed, function f can be defined as follows:

$$f(1) = g(1),$$

$$f(n + 1) = \min\Big\{g(m) \mid \big(g(m) \in \mathbb{N}\big) \wedge \big(b_{f(n)} \ll b_{g(m)}\big) \wedge \big(b_{g(n)} \ll b_{g(m)}\big)\Big\}.$$

Function f is total because the set whose minimum is the value of f is nonempty. This set is not empty because $b_{f(n)} \vee b_{g(n)} \ll x$ and because of Theorem 3.1.1, there is a $b \in B$ such that $b_{f(n)} \leqslant b_{f(n)} \vee b_{g(n)} \ll b \ll b \ll x$ and $b_{g(n)} \ll b_{f(n)} \vee b_{g(n)} \ll b \ll x$. In addition, function f is computable because the relation $\ll$ is recursively enumerable. Trivially, $b_{f(n)}$ is a $\ll$-chain that satisfies (4.11).

(ii)$\Rightarrow$(iii) and (iii)$\Rightarrow$(iv) Trivial.

(iv)$\Rightarrow$(i) Let g be a recursive function defined as follows:

$$g(1) = f(1),$$
$$g(n + 1) = \text{join}\big(g(n), f(n + 1)\big).$$

Since $b_{g(n)} = b_{f(1)} \vee \cdots \vee b_{f(n)}$, $(b_{g(n)})_{n\in\mathbb{N}}$ is a directed sequence such that $x = \bigvee_{n\in\mathbb{N}} b_{g(n)}$. Now, $b_m \ll x$ implies that there is a b such that $b_m \ll b \ll x$, that is, there is an $n \in \mathbb{N}$ such that $b_m \ll b_{g(n)}$ and so $b_m \ll x$ if and only if there an $n \in \mathbb{N}$ such that $b_m \ll b_{g(n)}$. Now, relation $\ll$ is recursively enumerable in B and so the set $\{m \in \mathbb{N} \mid b_m \ll x\}$ is recursively enumerable too. $\qquad\square$

In order to define the notion of decidability it is necessary to dualize the structures that have been used to define semidecidability. Given a lattice $(L, \leqslant)$, its dual or opposite is the structure $(L, \leqslant^{\text{op}})$ where $y \leqslant^{\text{op}} x$ if and only if $x \leqslant y$. Typically, the dual of a poset P is written as P^{op}. Assume that $(L, \leqslant)$ is a lattice and that $a, b \in L$. Then, a is *way above* b, denoted as $b \ll^{\text{op}} a$, when b is way below a in $(L, \leqslant^{\text{op}})$. Also, $b \ll^{\text{op}} a$ when, for every downward directed subset A of L, $\bigwedge A \leqslant b$ implies that there is an $x \in A$ such that $x \leqslant a$. Clearly,

$$b \ll^{\text{op}} a \Rightarrow a \leqslant b,$$

and, if L is a finite chain,

$$b \ll^{\text{op}} a \Leftrightarrow a \leqslant b.$$

When $L = [0, 1]$, then $b \ll^{\text{op}} a$ if and only if either $a = 1$ or $a < b$.

Definition 4.8.2 The quadruple $(L, \leqslant, B, \underline{B})$ is called an *effective above–below* lattice, or just *ab-lattice*, when the triples $(L, \leqslant, B)$ and $L, \leqslant^{\text{op}}, \underline{B})$ are based (effective) continuous lattices. The sequence $B = (b_n)_{n\in\mathbb{N}}$ is the *basis* and $\underline{B} = (\underline{b}_n)_{n\in\mathbb{N}}$ is the *opposite basis* of $(L, \leqslant, B, \underline{B})$.

The term *above–below* has been chosen because for all $a \in L$:

$$a = \bigvee\Big\{b \mid (b \in B) \wedge (b \ll a)\Big\} = \bigwedge\Big\{\underline{b} \mid (\underline{b} \in \underline{B}) \wedge (a \ll^{\text{op}} \underline{b})\Big\}.$$

This means that every element can be approximated both from below and from above. In addition, if this approximation process is effective, one gets the notion of decidable elements:

Definition 4.8.3 Assume that $(L, \leqslant, B, \underline{B})$ is an ab-lattice. Then, x is decidable if it is semidecidable in both $(L, \leqslant, B)$ and $(L, \leqslant^{\text{op}}, \underline{B})$, that is, if the sets

$$\Big\{n \mid (n \in \mathbb{N}) \wedge (b_n \ll a)\Big\} \text{ and } \Big\{n \mid (n \in \mathbb{N}) \wedge (x \ll^{\text{op}} \underline{b}_n)\Big\}$$

are recursively enumerable.

The next result follows from Proposition 4.8.1:

Proposition 4.8.2 *Assume that x is an element of an ab-lattice. Then, the following are equivalent:*

(i) *x is decidable;*

(ii) *there are two total recursive functions $h : \mathbb{N} \to \mathbb{N}$ and $k : \mathbb{N} \to \mathbb{N}$ such that $(b_{h(n)})_{n \in \mathbb{N}}$ preserves $\ll$ and $(\underline{b}_{-k(n)})_{n \in \mathbb{N}}$ reverse-preserves $\ll^{op}$ and*

$$\bigvee_{n \in \mathbb{N}} b_{h(n)} = x = \bigwedge_{n \in \mathbb{N}} \underline{b}_{-k(n)}; \tag{4.12}$$

(iii) *there are two recursive functions $h : \mathbb{N} \to \mathbb{N}$ and $k : \mathbb{N} \to \mathbb{N}$ such that $(b_{h(n)})_{n \in \mathbb{N}}$ is preserving $\ll$ and $(\underline{b}_{-k(n)})_{n \in \mathbb{N}}$ and (4.12) is satisfied; and*

(iv) *there is a nested effectively computable sequence $([b_{h(n)}, \underline{b}_{-k(n)}])_{n \in \mathbb{N}}$ of intervals such that*

$$\{x\} = \bigcap_{n \in \mathbb{N}} \left[b_{h(n)}, \underline{b}_{-k(n)} \right].$$

It is not difficult to construct ab-lattices by using an involution (i.e., negation operator) in L. Typically, for an involution $\neg : L \to L$, it holds that $\neg 0 = 1$, $\neg 1 = 0$, $\neg(x \wedge y) = \neg x \vee \neg y$, $\neg(x \vee y) = \neg x \wedge \neg y$, and $\neg\neg x = x$.

Definition 4.8.4 A structure $(L, \leqslant, \neg, B)$ is an effective lattice with an involution if $(L, \leqslant, B)$ is an effective lattice and $\neg$ is an involution such that

$$\left\{ (n, m) \mid \left((n, m) \in \mathbb{N} \times \mathbb{N} \right) \wedge \left(\neg b_n \ll \neg b_m \right) \right\}$$

is recursively enumerable.

Example 4.8.1 When $L = \{\lambda_0, \lambda_1, ..., \lambda_n\}$ is a finite chain, where $0 = \lambda_0 < \cdots < \lambda_n = 1$, then $\neg \lambda_i = \lambda_{n-i}$. Also, when $L = [0, 1]$, then $\neg \lambda = 1 - \lambda$.

An involution is actually an isomorphism between L and its opposite and since isomorphisms preserve relations, it holds that for any $x \in L$

$$x \ll^{op} y \Leftrightarrow \neg y \ll \neg x.$$

It is very easy to prove the following result:

Proposition 4.8.3 *Assume that $(L, \leqslant, \neg, B)$ is an effective lattice with an involution and that $\underline{B} = (\underline{b}_n)_{n \in \mathbb{N}}$, where $\underline{b}_n = \neg b_n$. Then, $(L, \leqslant, B, \underline{B})$ is an effective ab-lattice. In addition, x is decidable if and only if both x and $\neg x$ are semidecidable.*

The unit interval is an effective ab-lattice where $B = \underline{B} = \bar{\text{I}}$. In this case, an x is decidable when the sets $\{r \in \bar{\text{I}} \mid r < x\}$ and $\{r \in \bar{\text{I}} \mid r > x\}$ are both recursively enumerable, which means that x is a recursive real number.

Let X be a nonempty set. Then the set L^X, where L is a lattice, is the set of all L-subsets of X. In different words, the set L^X is the set of all functions with domain the set X and codomain the set L. In Sect. 3.3 there is a description of the basic operations between elements of the set L^X. Suppose that $A \in L^X$. Then,

$$\text{Supp}(A) = \left\{ x \mid (x \in X) \wedge (A(x) \neq 0) \right\},$$

$$\text{Cosp}(A) = \left\{ x \mid (x \in X) \wedge (A(x) \neq 1) \right\}.$$

The subset A is called finite or co-finite when the sets $\text{Supp}(A)$ and $\text{Cosp}(A)$ are finite, respectively. In addition, both the empty set and A are considered to be finite sets. The classes of finite and co-finite L-subsets are written as $\text{Fin}(L^X$ and $\text{Cof}(L^X)$, respectively.

4.8.2   Effective Lattices of L-Fuzzy Subsets

Although the theory presented so far seems quite interesting, there is no direct link to fuzzy subset theory or, more generally, to L-fuzzy subset theory. The theorem that follows provides the required link:

Theorem 4.8.1 *Suppose that $(L, \leqslant, B)$ is an effective lattice. Then, the class L^X of L-subsets of X is an effective lattice having as basis the class $\text{Fin}(B^S)$ of finite L-subsets of X with values in B. Also, for every $S_1, S_2 \in L^X$, $S_1 \ll S_2$ if and only if S_1 is finite and $S_1(x) \ll S_2(x)$ for all $x \in X$.*

Remark 4.8.1 An L-subset A is semidecidable provided that the set

$$\left\{ n \mid (n \in \mathbb{N}) \wedge (b_n \ll A) \right\} = \left\{ n \mid (n \in \mathbb{N}) \wedge (\forall i \in \text{Supp}(b_n) \Rightarrow b_n(i) \ll A(i)) \right\}$$

is recursively enumerable.

The proof of the following result is an immediate consequence of Proposition 4.8.1.

Theorem 4.8.2 *Assume that $(L, \leqslant, B)$ is an effective lattice and that $A \in L^X$. Then, the following statements are equivalent:*

(i) *A is semidecidable;*

(ii) *there is a recursive function $h : X \times \mathbb{N} \to B$ that is $\ll$-increasing with respect to n and such that*

$$A(x) = \bigvee_{n \in \mathbb{N}} h(x, n); \text{ and}$$

(iii) *there is a recursive function $h : X \times \mathbb{N} \to B$ that is increasing with respect to n and such that*

$$A(x) = \bigvee_{n \in \mathbb{N}} h(x, n).$$

The next proposition will be used to define the notion of a decidable L-subset.

Proposition 4.8.4 *Assume that $(L, \leqslant, B, \underline{B})$ is an ab-lattice. Then, L^X is an effective ab-lattice with opposite basis the class $\mathrm{Cof}(B^X)$ of co-finite L-subsets of X with values in B. If $(L, \leqslant, \neg, B)$ is an effective lattice with an involution, then L^X is an effective lattice with the complement as an involution.*

Remark 4.8.2 Suppose that $(L, \leqslant, B, \underline{B})$ is an ab-lattice. Then, an element A is decidable if and only if both A and its complement $\neg A$ are semidecidable.

Definition 4.8.5 Assume that A_1 and A_2 are two L-fuzzy subsets. Then, A_1 is m-reducible to A_2, in symbols $A_1 \leqslant_m A_2$, when there is a recursive function $h : X \to X$ such that for any $x \in X$, $A_1(x) = A_2(h(x))$.

4.8.3   Chains and the Unit Interval

Up till now, it was assumed that L can be any lattice, but in this section it will be assumed that L is either a (finite) chain or the unit interval.

Proposition 4.8.5 *Assume that L is a finite chain. Then the class L^X of L-subsets if X is an effective lattice. In addition, the complement can be seen as an involution which makes it an effective ab-lattice. Its basis is the class $\mathrm{Fin}(L^X)$ of finite L-subsets of X, its opposite basis is the class $\mathrm{Cof}(L^X)$ of co-finite L-subsets of X. Also,*

$$A_1 \ll A_2 \Leftrightarrow A_1 \subseteq A_2 \text{ and } A_1 \in \mathrm{Fin}(L^X)$$

and

$$A_1 \ll^{\mathrm{op}} A_2 \Leftrightarrow A_1 \subseteq A_2 \text{ and } A_2 \in \mathrm{Cof}(L^X).$$

Remark 4.8.3 The class $\wp(X)$ of subsets of X is an effective lattice with an involution whose basis and opposite basis are the classes of finite and co-finite subsets of X, respectively. In addition,

$$X_1 \ll X_2 \Leftrightarrow X_1 \subseteq X_2 \text{ and } X_1 \in \mathrm{Fin}(X)$$

and

$$X_1 \ll^{\mathrm{op}} X_2 \Leftrightarrow X_1 \subseteq X_2 \text{ and } X_2 \in \mathrm{Cof}(X).$$

Also, the proposed notions of decidability and semidecidability coincide with the known ones.

Proposition 4.8.6 *Suppose that L is finite chain. Then, an L-subset A is semidecidable if and only if there exists a recursive function $h : X \times \mathbb{N} \to L$, which is increasing with respect to the second argument, such that*

$$A(x) = \bigvee_{n \in \mathbb{N}} h(x, n).$$

In addition, A is decidable if and only if it is computable.

The important question is whether this approach to semidecidability in the case of finite chains is the only possible extension of the ordinary notion of semidecidability that has the following properties:

- the constant L-subsets are semidecidable;

- the union of two semidecidable L-subsets is semidecidable; and

- the intersection of two semidecidable L-subsets is semidecidable.

This can be shown using the following equation, which holds for the α-cuts of some L-subset A:

$$A(x) = \bigcup_{\lambda \in L} \lambda \wedge {}^\alpha A.$$

This shows that the lattice of the L-subsets is the lattice generated by the constant L-subsets and the crisp L-subsets.

Proposition 4.8.7 *Suppose that L is a finite chain. Then, the following statements are equivalent:*

(i) A is a semidecidable L-subset; and

(ii) all α-cuts of A are recursively enumerable.

In addition, the lattice of the semidecidable L-subsets is the lattice generated by the recursively enumerable subsets and the constant L-subsets.

Proposition 4.8.8 *Suppose that L = [0,1]. Then, the class of L-subsets of X (i.e., the class of fuzzy subsets of X) is an effective lattice where the complement is its involution. Its basis is the class $\mathrm{Fin}\left(\bar{I}^X\right)$ and its opposite basis is the class $\mathrm{Cof}\left(\bar{I}^X\right)$. In addition,*

$$A_1 \ll A_2 \Leftrightarrow A_1 \in \mathrm{Fin}\left(\bar{I}^X\right) \text{ and } A_1(x) < A_2(x), \forall x \in \mathrm{Supp}(A_1)$$

and

$$A_1 \ll^{\mathrm{op}} A_2 \Leftrightarrow A_2 \in \mathrm{Cof}\left(\bar{I}^X\right) \text{ and } A_1(x) < A_2(x), \forall x \in \mathrm{Cosp}(A_1).$$

When L = [0,1], Proposition 4.8.7 becomes meaningless because the α-cut of a semidecidable L-subset is not necessarily recursively enumerable. Instead, one has to rely on the following result, whose proof relies on a number of results that can be found in the literature (e.g., see [62]).

Theorem 4.8.3 *A subset of X is an α-cut of a semidecidable fuzzy subset if and only if it is a Σ_2^0 set.*

4.8.4   Relationship to L-Languages and L-Turing Machines

When the plausibility degrees associated with "fuzzy" grammars and languages are drawn from some lattice L, the resulting structures are called L-grammars and L-languages, respectively.

Theorem 4.8.4 *Assume that $L(G)$ is an L-language either generated by an L-grammar G or accepted by an L-Turing machine. Then, $L(G)$ is a semidecidable L-subset.*

Proof Suppose that $L(G)$ is generated by an L-grammar and so it satisfies an equation analogous to (4.1) on p. 53. Note that here max should be replaced by a suitable conjunction operator $\otimes$ that has to be order-preserving, associative, commutative, and such that $x \otimes 1 = x$ for all $x \in L$. Then, $L(G)$ is semidecidable because for any string ξ it is possible to enumerate in an effective way the class of derivations of ξ. The second part of the theorem can been shown by employing a similar argument. ☐

The following theorem is interesting because it states that three "different" approaches to effectiveness are the same.

Theorem 4.8.5 *Suppose that L is a finite chain and that A is an L-subset. Then, the following are equivalent:*

- *A is semidecidable;*

- *there is an L-grammar that generates A; and*

- *there is an L-Turing machine that accepts A.*

Proof Suppose that L is a finite chain whose elements are $0 = \lambda_0 < \cdots < \lambda_n = 1$ and that A is semidecidable. Then, all the α-cuts $^{\lambda_i}A$ are recursively enumerable. For every $0 < i \leq n$, assume that $G_i = (V_N, V_T, P_i, S)$ is a grammar that generates $^{\lambda_i}A$. Let $G = (V_N, V_T, P, S)$ be the grammar created by setting

$$P(\alpha \xrightarrow{\rho} \beta) = \bigvee \left\{ \lambda_i \,\middle|\, (\alpha \xrightarrow{\rho} \beta) \in P_i \right\},$$

where here $\otimes$ is the minimum operator. Then, it is easy to see that the L-language accepted by a L-Turing machine is actually the L-subset A.

A similar argument can be employed to prove the case for L-Turing machines and the reader should have no difficulty to complete the proof. ☐

Wiedermann [141] proves a series of results that can be summarized as follows:

Proposition 4.8.9 *Assume that $(M, \otimes, 1)$ is a finitely generated sub-monoid of $(L, \otimes, 1)$. Then, every nonempty subset of M contains a maximal element, that is, if $F \subseteq M$, there is an element $a \in F$ such that $x \preccurlyeq a$ for all $x \in F$.*

A direct consequence of the previous proposition is that the join operator in the equations related to languages and grammars can safely be replaced by the max operator. The following result is about the properties of L-languages generated by L-grammars accepted by L-Turing machines.

Theorem 4.8.6 *Suppose that L is totally ordered and that A is an L-language generated by an L-grammar that is accepted by an L-Turing machine. Then, the values assumed by A are in B and the α-cuts $^\lambda A$, where $\lambda \in B$, are recursively enumerable.*

Proof Assume that A is generated by an L-grammar and because of the previous proposition

$$A(w) = \max\Big\{\lambda(\pi) \,\Big|\, \pi \text{ is a derivation chain of } w\Big\}.$$

Also,

$$^\lambda A = \Big\{x \,\Big|\, \big(x \in X\big) \wedge \big(\text{there is a derivation chain } \pi \text{ such that } \lambda(\pi) \geq \lambda\big)\Big\}.$$

In order to prove that $^\lambda A$ is recursively enumerable, it is sufficient to note that the function $\lambda(w)$ is effectively computable and that the relation $\lambda(\pi) \geq \lambda$ is decidable.

In the case of L-Turing machines, a similar argument can be employed. □

A side effect of this theorem is that Theorem 4.8.5 is not valid when L is an infinite chain. For instance, when $L = [0, 1]$, then every semidecidable L-language that assumes irrational values is an example of semidecidable L-subset for which there is no generating L-grammar and accepting L-Turing machine. In this respect, the following result is also very interesting.

Theorem 4.8.7 *Let L be the effective lattice defined in the unit interval and let* big $: I^+ \to$ *$[0, 1]$ be the fuzzy subset of all "big words" defined as follows:*

$$\text{big}(w) = 1 - 1/\text{length}(w)$$

for every word w. Then big *is a decidable L-language such that*

- big *has only rational values;*
- *the α-cuts of* big *are all decidable;*
- *there is no L-grammar able to generate* big; *and*
- *there is no L-Turing machine able to accept* big.

Proof Just note that there is no maximum in the range of function big and so no L-grammar can generate big and also no L-Turing machine can accept big. □

It is quite interesting that no L-Turing machine and no L-Grammar can accept or generate function big, respectively, whereas it is a decidable function. Gerla provides a remedy for this problem by stating a different version of the soft Church–Turing thesis and then by demanding revised definitions of fuzzy Turing machines and fuzzy grammars. However, it makes no sense to extend the soft Church–Turing thesis—it would be far better to admit that the soft thesis poses an artificially computational barrier. In different words, by accepting that there are "solvable" problems that cannot be solved by fuzzy Turing machines, one implicitly admits that there are computing devices capable to solve these problems. That is, hypercomputation by definition!

4.9   Fuzzy Complexity Theory

Wiedermann [142] has not only investigated the computational power of fuzzy Turing machines but also their computational efficiency. First, one needs to define what a bounded computation is.

Definition 4.9.1 Assume that $\mathscr{F}$ is classical fuzzy nondeterministic Turing machine and that $L(\mathscr{F})$ is the language decided by $\mathscr{F}$. Then, $\mathscr{F}$ is of time complexity $T(n)$ if for all $n \in \mathbb{N}$ and all inputs of length n, machine $\mathscr{F}$ accepts all $L(\mathscr{F})$ words of length n in at most $T(n)$ steps.

The next step involves the definition of the equivalent of the ordinary NP class (see Sect. 2.7) in the fuzzy setting:

Definition 4.9.2 The class of all fuzzy languages accepted by a classical fuzzy nondeterministic Turing machine in polynomial time (i.e., when $T(n)$ is a polynomial) is denoted Fuzzy-NP.

The following theorem characterizes the computational efficiency of fuzzy Turing machines:

Theorem 4.9.1 *Fuzzy-NP = $NP \cup$ co-NP = $\Sigma_1^P \cup \Pi_1^P$.*

   Proof (Sketch) The proof is similar to proof of Theorem 4.4.1 on p. 76. In particular, Fuzzy-NP corresponds to Φ, NP to Σ_1^0, and co-NP to Π_1^0. In addition, it is necessary to explicitly state that all machines involved in simulations are polynomial bounded, something clearly obvious.   □

5. Other Fuzzy Models of Computation

Fuzzy Turing machines are not the only way to perform computation in a vague environment. Other models of fuzzy computation are inspired by biological phenomena or, more generally, by natural phenomena. Fuzzy P systems and the fuzzy chemical abstract machine are such models of computation. Some of these models have been studied in some detail while others are just emerging proposals.

5.1 Fuzzy P Systems

P systems is a model of computation that mimics the way cells live and function. The model was introduced by Gheorghe Păun [100] (see also [101] for a detailed but rather outdated presentation of the field). Typically, a cell has a membrane that surrounds it, separates its interior from its environment, regulates what goes in and out, etc. Inside the membrane, the cytoplasm takes up most of the cell volume. Various organelles (i.e., specialized subunits within a cell that have a specific function) are "floating" inside the cytoplasm. Just like a cell, a P system is enveloped in a porous membrane that allows objects to move in and out. Inside a P system there is an indefinite number of nested compartments, that is, compartments that may contain other compartments, etc., each of them enveloped by a porous membrane. Also, there is a designated compartment called the *output compartment*. In addition, each compartment may contain "solid," possibly repeated objects, that is, a multiset of objects, while it is associated with a set of multiset rewriting rules. The system operates in discrete time and these rules specify what changes can possible happen inside a compartment at each tick of the clock. In general, compartments cannot be deleted while objects may be multiplied, deleted, or introduced in a compartment. Computation stops when no rule is applicable and the result of the computation equals the number of objects that have been accumulated in the output compartment.

As they stand, P systems provide no real insight into either computability theory or complexity theory. For instance, it was claimed that P systems have at most the computational power of Turing machines (see [101]). However, by properly extending the definition of P systems, one may get new systems with interesting properties. For instance, by fuzzifying the data manipulated by a P system, the result is a new structure that is in principle capable of computing any positive real number. Also, by replacing the calm and quiet environment that surrounds ordinary P systems with an active environment that continuously and reciprocally affects membrane structures, one gets a truly interactive systems.

P systems with fuzzy data were first defined by this author [124] and they were generalized again by the same author [129, 130]. These systems manipulate generalizations of the

A. Syropoulos, *Theory of Fuzzy Computation*, IFSR International Series on Systems Science and Engineering 31, DOI 10.1007/978-1-4614-8379-3_5,

concept of a multiset. Although multisets are widely used in computer science, still not everyone is familiar with the relevant notions, which are introduced in the next paragraph.

Definition 5.1.1 Given some universe X, a multiset A that draws elements from X may contain multiply copies of any element $x \in X$. In particular, a multiset A is characterized by a function $A : X \to \mathbb{N}$, where $A(x) = n$ denotes that A contains n copies of x.

The various operations between multisets are an obvious extension of the usual set operations.

Definition 5.1.2 Assume that $A, B : X \to \mathbb{N}$ are two multisets. Then, the various operations are defined as follows:

Support The *support* of A is a set S whose characteristic function is

$$\chi_S(x) = \min\{A(x), 1\}.$$

Subsethood A is a subset of B provided that $A(x) \le B(x)$ for all $x \in X$.

Cardinality The cardinality of A equals

$$\sum_{x \in X} A(x).$$

Sum The sum of A and B, denoted as $A \uplus B$, is a multiset such that for all $x \in X$,

$$(A \uplus B)(x) = A(x) + B(x).$$

Union The union of A and B, denoted as $A \cup B$, is a multiset such that for all $x \in X$,

$$(A \cup B)(x) = \max\{A(x), B(x)\}.$$

Intersection The intersection of A and B, denoted as $A \cap B$, is a multiset such that for all $x \in X$,

$$(A \cap B)(x) = \min\{A(x), B(x)\}.$$

Assume that A is a multiset. Then, a "fuzzy multisubset" of A should be a structure that contains the elements of A to some degree. To avoid confusion with other structures, these structures have been dubbed *multi-fuzzy sets* and have been defined as follows [125]:

Definition 5.1.3 Assume that $M : X \to \mathbb{N}$ characterizes a multiset M. Then, a multi-fuzzy subset of M is a structure A that is characterized by a function $A : X \to \mathbb{N} \times [0, 1]$ such that if $M(x) = n$, then $A(x) = (n, i)$. In addition, the expression $A(x) = (n, i)$ denotes that the degree to which each of the n copies of x belong to A is i.

Given a multi-fuzzy set A, one can define the following two functions: the *multiplicity* function $A_\pi : X \to \mathbb{N}$ and the *membership* function $A_\sigma : X \to [0, 1]$. If $A(x) = (n, i)$, then $A_\pi(x) = n$ and $A_\sigma(x) = i$.

Definition 5.1.4 Suppose that A is a multi-fuzzy set having the set X as its universe. Then, its cardinality, denoted as card A, equals

$$\sum_{a \in X} A_\pi(a) A_\sigma(a).$$

In order to give a formal definition of a P system, it is required to formally define membrane structures:

Definition 5.1.5 Let $V = \{[,]\}$ be an alphabet. Then, the set MS is the least set inductively defined as follows:

(i) $[] \in$ MS; and

(ii) if $\mu_1, \mu_2, \dots \mu_n \in$ MS, then $[\mu_1 \dots \mu_n] \in$ MS.

With these preliminary definitions, let me proceed with the formal definition of P systems with fuzzy data.

Definition 5.1.6 A P system with fuzzy data is a construction

$$\Pi_{\text{FD}} = (O, \mu, w^{(1)}, \dots, w^{(m)}, R_1, \dots, R_m, i_0),$$

where

(i) O is an alphabet (i.e., a set of distinct entities) whose elements are called *objects*.

(ii) μ is the membrane structure of degree $m \geq 1$, which is the depth of the corresponding tree structure; membranes are injectively labeled with successive natural numbers starting with one.

(iii) Each $w^{(i)} : O \to \mathbb{N} \times I, 1 \leq i \leq m$ is a multi-fuzzy set over O associated with the region surrounded by membrane i.

(iv) $R_i, 1 \leq i \leq m$, are finite sets of multiset rewriting rules (called *evolution rules*) over O. An evolution rule is of the form $u \to v, u \in O^*$ and $v \in O^*_{\text{TAR}}$, where $O_{\text{TAR}} = O \times \text{TAR}$,

$$\text{TAR} = \{\text{here}, \text{out}\} \cup \{\text{in}_j \mid 1 \leq j \leq m\}.$$

The keywords "here," "out," and "in$_j$" are used to specify the current compartment (i.e., the compartment the rule is associated with), the compartments that surrounds the current compartment, and the compartment with label j, respectively. The effect of each rule is the removal of the elements of the left-hand side of the rule from the current compartment (i.e., elements that match the left-hand side of a rule are removed from the current compartment) and the introduction of the elements of the right-hand side to the designated compartments. Also, the rules implicitly transfer the fuzzy degrees to membership in their new "home set."

(v) $i_0 \in \{1, 2, \dots, m\}$ is the label of an elementary membrane (i.e., a membrane that does not contain any other membrane), called the *output* membrane.

Before discussing the main result concerning the computational power of these systems let me say a few things about the simple P system with fuzzy data that is depicted in Fig. 5.1.

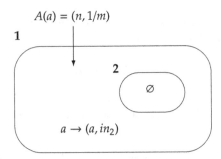

Figure 5.1: A simple P system with fuzzy data

Initially, this P system contains n objects to degree equal to $1/m$ in compartment 1. Once computation commences, these objects will be transferred into the output compartment 2, which is initially empty. The result of the computation (i.e., the cardinality of the multi-fuzzy set contained in compartment 2) is equal to n/m. Thus, the result of this particular computation is a positive rational number.

The really interesting thing about these systems is that I have not managed to find any limits on what can be actually computed. Technically, a real number $x \in \mathbb{R}$, where $\mathbb{R}$ is the set of real numbers, is called *computable* if there is a computable sequence $(r_n)_{n \in \mathbb{N}}$ of rational numbers which converges to x effectively, that is, for all $n \in \mathbb{N}$, $|x - r_n| < 2^{-n}$ (see [140, 153] for details). In other words, not all real numbers are *computable*. However, one should not forget that the definition of *computability* is hard-wired to the computational capabilities of the Turing machine. The crucial question is whether there are any limits that prohibit the computation of certain numbers with fuzzy P systems? It seems that these systems go beyond the Church–Turing barrier because their set of input values is drastically larger than that of the Turing machine. In addition, numbers are represented by cardinalities, which is rather different from the Turing machine representation of results. However, it is an open problem the determination of the exact computational power of these systems.

It is not difficult to define L-multi-fuzzy subsets and L-fuzzy hybrid subsets. However, one needs to carefully define the various set operations. L-multi-fuzzy sets are an extension of multi-fuzzy sets just like L-fuzzy sets are an extension of fuzzy sets.

Let me now present the definitions of union, intersection, and sum of L-multi-fuzzy sets:

Definition 5.1.7 Assuming that $A, B : X \to L \times \mathbb{N}$ are two L-multi-fuzzy sets. Then, their union, denoted as $A \cup B$, is defined as follows:

$$\left(A \cup B\right)(x) = \left(A_\mu(x) \vee B_\mu(x), \max\left\{A_m(x), B_m(x)\right\}\right).$$

Definition 5.1.8 Assuming that $A, B : X \to L \times \mathbb{N}$ are two L-multi-fuzzy sets. Then, their intersection, denoted as $A \cap B$, is defined as follows:

$$\left(A \cap B\right)(x) = \left(A_\mu(x) \wedge B_\mu(x), \min\left\{A_m(x), B_m(x)\right\}\right).$$

Definition 5.1.9 Suppose that $A, B : X \rightarrow L \times \mathbb{N}$ are two L-multi-fuzzy sets. Then, their sum, denoted as $A \uplus B$, is defined as follows:

$$(A \uplus B)(x) = \Big(A_\mu(x) \vee B_\mu(x), A_m(x) + B_m(x)\Big).$$

Although it is crystal clear, it is necessary to say that $\vee$ and $\wedge$ are operators that are part of the definition of the frame L. And as such they have a number of properties (e.g., they are idempotent; see [137, p. 15] for details) that, naturally, affect the properties of the operations defined so far. Indeed, these operations have the following properties:

Theorem 5.1.1 *For any three L-multi-fuzzy sets* $A, B, C : X \rightarrow L \times \mathbb{N}$ *the following equalities hold:*

(i) *Commutativity:*

$$
\begin{aligned}
A \cup B &= B \cup A \\
A \cap B &= B \cap A \\
A \uplus B &= B \uplus A
\end{aligned}
$$

(ii) *Associativity:*

$$
\begin{aligned}
A \cup (B \cup C) &= (A \cup B) \cup C \\
A \cap (B \cap C) &= (A \cap B) \cap C \\
A \uplus (B \uplus C) &= (A \uplus B) \uplus C
\end{aligned}
$$

(iii) *Idempotency:*

$$
\begin{aligned}
A \cup A &= A \\
A \cap A &= A
\end{aligned}
$$

(iv) *Distributivity:*

$$
\begin{aligned}
A \cap (B \cup C) &= (A \cap B) \cup (A \cap C) \\
A \cup (B \cap C) &= (A \cup B) \cap (A \cup C)
\end{aligned}
$$

(v) *Distributivity of sum:*

$$
\begin{aligned}
A \uplus (B \cup C) &= (A \uplus B) \cup (A \uplus C) \\
A \uplus (B \cap C) &= (A \uplus B) \cap (A \uplus C)
\end{aligned}
$$

Proof

(i) Although this is easy, I will prove all cases:

$$(A \cup B)(z) = \left(A_\mu(z) \vee B_\mu(z), \max\left\{A_m(z), B_m(z)\right\}\right)$$

$$= \left(B_\mu(z) \vee A_\mu(z), \max\left\{B_m(z), A_m(z)\right\}\right)$$

$$= (B \cup A)(z)$$

$$(A \cap B)(z) = \left(A_\mu(z) \wedge B_\mu(z), \min\left\{A_m(z), B_m(z)\right\}\right)$$

$$= \left(B_\mu(z) \wedge A_\mu(z), \min\left\{B_m(z), A_m(z)\right\}\right)$$

$$= (B \cap A)(z)$$

$$(A \uplus B)(z) = \left(A_\mu(z) \vee B_\mu(z), A_m(z) + B_m(z)\right)$$

$$= \left(B_\mu(z) \vee A_\mu(z), B_m(z) + A_m(z)\right)$$

$$= (B \uplus A)(z)$$

(ii) I will prove only the first case as the others can be proved similarly:

$$(A \cup (B \cup C))(z) = \left(A_\mu(z) \vee \left(B_\mu(z) \vee C_\mu(z)\right), \max\left\{A_m(z), \max\left\{B_m(z), C_m\right\}\right\}\right)$$

$$= \left(\left(A_\mu(z) \vee B_\mu(z)\right) \vee C_\mu(z), \max\left\{\max\left\{A_m(z), B_m(z)\right\}, C_m\right\}\right)$$

$$= ((A \cup B) \cup C)(z)$$

(iii) As in the previous case, I will prove only the first case as the other can be proved similarly:

$$(A \cup A)(z) = \left(A_\mu(z) \vee A_\mu(z), \max\{A_m(z), A_m(z)\}\right)$$

$$= \left(A_\mu(z), A_m(z)\right)$$

$$= A(z)$$

(iv) The proof of this case follows from the fact that the following equalities are true for the any three elements of a frame:

$$x \wedge (y \vee z) = (x \wedge y) \vee (x \wedge y),$$
$$x \vee (y \wedge z) = (x \vee y) \wedge (x \vee y).$$

(v) As with the previous case the proof for this case follows from the fact that for any $x, y, z \in \mathbb{N}$ the following equalities hold:

$$x + \max\{y, z\} = \max\{x + y, x + z\},$$
$$x + \min\{y, z\} = \min\{x + y, x + z\}.$$

$\square$

The α-cut of a fuzzy subset is just a crisp set. Similarly, the α-cut of an L-multi-fuzzy set has to be a multiset. Indeed, the following definition is in the spirit of the general theory of fuzzy sets:

Definition 5.1.10 Suppose that A is an L-multi-fuzzy set with universe the set X and that $\alpha \in L$. Then the α-cut of A, denoted by $^\alpha A$, is the multiset

$$^\alpha A = \bigcup_{\substack{x \in X \\ \alpha \leq A_\mu(x)}} [x]_{A_m(x)},$$

where $[x]_n$ denotes a multiset that consists of exactly n copies of x only.

Not so surprisingly, the properties of the α-cut of L-multi-fuzzy sets are similar to those of plain fuzzy sets. These properties are summarized below:

Theorem 5.1.2 *Assume that A and B are two L-multi-fuzzy sets with universe the set X. Then the following properties hold:*

(i) *if $\alpha \leq \beta$, then $^\alpha A \supseteq {}^\beta A$; and*

(ii) $^\alpha(A \cap B) = {}^\alpha A \cap {}^\alpha B$, $^\alpha(A \cup B) = {}^\alpha A \cup {}^\alpha B$, *and* $^\alpha(A \uplus B) = {}^\alpha A \uplus {}^\alpha B$.

Proof

(i) Let $x \in X$ and $\alpha \leq \beta$. If $A_\mu(x) \nleq \beta$, then $^\alpha A(x) = {}^\beta A(x)$. If $\alpha \leq A_\mu(x) \leq \beta$, then $^\alpha A(x) \geq {}^\beta A(x)$. If $\alpha \nleq A_\mu(x)$, then $^\alpha A(x) = {}^\beta A(x) = 0$. Thus, for all possible cases $^\alpha A(x) \geq {}^\beta A(x)$, which means that $^\alpha A \supseteq {}^\beta A(x)$.

(ii) I will prove only the first case; the others can be proved similarly. Assume that $^\alpha(A \cap B) = n$. Then, this means that

$$\min\left\{ A_m(x), B_m(x) \right\} = n.$$

Also, it implies that $(A \cap B)_\mu(x) \leq \alpha$ and, hence, $A_\mu(x) \wedge B_\mu(x) \leq \alpha$. From this, one can immediately deduce that $A_\mu(x) \geq \alpha$ and $B_\mu(x) \geq \alpha$. Suppose now that $A_m(x) = n_1$ and $B_m(x) = n_2$. Then this means that $^\alpha A(x) = n_1$ and $^\alpha B(x) = n_2$ and so

$$\min\left\{ ^\alpha A(x), {}^\alpha B(x) \right\} = n.$$

$\square$

Hybrid sets and *new* sets are generalization of multisets and sets, respectively. In a hybrid set the multiplicity of an element can be either a negative number, zero, or a positive number. An element may belong one, zero, or minus one times to a new set. Thus, a new set is to a hybrid set what is a set to a multiset, that is, a special case. Let me now give the definition of the hybrid set due to Daniel Loeb [89]:

Definition 5.1.11 Given a universe X, any function $f : X \rightarrow \mathbb{Z}$, where $\mathbb{Z}$ is the set of all integers, characterizes a hybrid set. The value of $f(x)$ is said to be the multiplicity of the element x. If $f(x) \neq 0$, then x is a member of f and this is denoted by $x \in f$. Otherwise, it is not a member of f and this is denoted by $x \notin f$. The cardinality of the hybrid set f equals

$$\sum_{x \in X} f(u).$$

The definition of L-fuzzy hybrid sets follows immediately:

Definition 5.1.12 An L-fuzzy hybrid set A is a mathematical structure that is characterized by a function $A : X \rightarrow L \times \mathbb{Z}$, where L is a frame, and it is associated with a L-fuzzy set $A' : X \rightarrow L$. More specifically, the equality $A(x) = (\ell, n)$ means that A contains exactly n copies of x and $A'(x) = \ell$.

As in the case of multi-fuzzy sets, one can define for a L-fuzzy hybrid set the following two functions: the multiplicity function $A_m : X \rightarrow \mathbb{Z}$ and the membership function $A_\mu : X \rightarrow L$. Clearly, if $A(x) = (\ell, n)$, then $A_m(x) = n$ and $A_\mu(x) = \ell$. Notice that it is equally easy to define the corresponding functions for an L-multi-fuzzy set. Let me now define the cardinality of an L-fuzzy hybrid set:

Definition 5.1.13 Assume that A is an L-fuzzy hybrid set that draws elements from a universe X. Then its cardinality equals

$$\sum_{x \in X} A_\mu(x) \otimes A_m(x),$$

where $\otimes : L \times \mathbb{Z} \rightarrow \mathbb{R}$ is a binary multiplication operator that is used to compute the product of $\ell \in L$ times $n \in \mathbb{Z}$.

Example 5.1.1 If $L = [0,1] \times [0,1]$ (i.e., when extending "intuitionistic" fuzzy sets,[1] see [128]), then $(i, j) \otimes n = in - jn$.

Remark 5.1.1 When $L = [0,1]$, then $\otimes$ is the usual multiplication operator.

The cardinality of a set is equal to the number of elements the set contains. Clearly, the previous definition is not in spirit with this assumption. However, hybrid sets may contain elements that occur a negative number of times. Thus, one may think that we should take this fact under consideration when computing the cardinality of a hybrid set and, more generally, the cardinality of an L-fuzzy hybrid set. So, it makes sense to introduce the notion of a *strong* cardinality defined as follows:

---

1. "Intuitionistic" fuzzy sets were introduced by Atanassov [4]. In a nutshell, every element belongs to an intuitionistic fuzzy set to a degree equal to μ and does not belong to it to a degree equal to ν while $\mu \neq 1 - \nu$ and $0 \leq \mu + \nu \leq 1$. In general, the term "intuitionistic" is considered as a misnomer.

Definition 5.1.14 Assume that A is an L-fuzzy hybrid set that draws elements from a universe X. Then, its *strong* cardinality is defined as follows:

$$\text{card } A = \sum_{x \in X} A_\mu(x) \otimes |A_m(x)|,$$

where $|A_m(x)|$ denotes the absolute value of $A_m(x)$.

In order to complete the presentation of the basic properties of fuzzy hybrid sets, it is necessary to define the notion of subsethood. Before, going on with this definition, I will introduce the (new) partial order $\lesssim$ over $\mathbb{Z}$. In particular, if $n, m \in \mathbb{Z}$, then

$$n \lesssim m \equiv (n = 0) \vee$$
$$\left((n > 0) \wedge (m > 0) \wedge (n \leq m) \right) \vee$$
$$\left((n < 0) \wedge (m > 0) \right) \vee$$
$$(|n| \leq |m|).$$

Note that here $\wedge$ and $\vee$ denote the classical logical conjunction and disjunction operators, respectively. In addition, the symbols $\leq$ and $<$ are the well-known ordering operators, and $|n|$ is the absolute value of n.

Example 5.1.2 From the previous definition it should be obvious that $0 \lesssim n$, for all $n \in \mathbb{Z}$. Also, $3 \lesssim 4$, $-3 \lesssim 4$, and $-4 \lesssim -3$.

But what kind of structure is the pair $(\mathbb{Z}, \lesssim)$? The answer is easy with the help of the following result:

Proposition 5.1.1 *The relation $\lesssim$ is a partial order.*

Proof I have to prove that the relation $\lesssim$ is reflexive, antisymmetric, and transitive:

Reflexivity Assume that $a \in \mathbb{Z}$. Then, if $a = 0$, $a \lesssim a$ from the first part of the disjunction. If $a < 0$, then $a \lesssim a$ from the fourth part of the disjunction, and if $a > 0$, then $a \lesssim a$ from the second part of the disjunction.

Antisymmetry Assume that $a, b \in \mathbb{Z}$, $a \lesssim b$, and $b \lesssim a$. Then, if $a = 0$ this implies that $b = 0$ and so $a = b$. If $a < 0$, then it follows that $b < 0$, $|a| \leq |b|$, and $|b| \leq |a|$, which implies that $a = b$. Similarly, if $a > 0$, then it follows that $b > 0$, $a \leq b$, and $b \leq a$, which implies that $a = b$.

Transitivity Assume that $a, b, c \in \mathbb{Z}$, $a \lesssim b$, and $b \lesssim c$. Then, if $a = 0$, then clearly $a \lesssim c$. If $a < 0$ and $b < 0$, then either $c < 0$ or $c > 0$, but since $|b| \leq |c|$, this implies that $a \lesssim c$. If $a > 0$ and $b > 0$, then $c > 0$ and since $b \leq c$ this implies that $a \lesssim c$. If $a < 0$ and $b > 0$, then since $b \lesssim c$, this implies that $c > 0$, which means that $a \lesssim c$.

□

Let us now proceed with the definition of the notion of subsethood for L-fuzzy hybrid sets:

Definition 5.1.15 Assume that $A, B : X \to L \times \mathbb{Z}$ are two L-fuzzy hybrid sets. Then, $A \subseteq B$ if and only if $A_\mu(x) \leq B_\mu(x)$ and $A_m(x) \lesssim B_m(x)$ for all $x \in X$.

Remark that for all $\ell_1, \ell_2 \in L$, $\ell_1 \leq \ell_2$ if ℓ_1 is "less than or equal" to ℓ_2 in the sense of the partial order defined over L. The definition of subsethood for L-multi-fuzzy sets is more straightforward:

Definition 5.1.16 Assume that $A, B : X \to L \times \mathbb{N}$ are two L-multi-fuzzy sets. Then, $A \subseteq B$ if and only if $A_\mu(x) \leq B_\mu(x)$ and $A_m(x) \leq B_m(x)$ for all $x \in X$.

Loeb has shown that the set of all subsets of a given hybrid set with the subsethood relation do not form a lattice. This means that if f and g are two hybrid sets, then if they have lower bounds, they do not necessarily have a greatest lower bound. Similarly, if f and g have upper bounds, then they do not necessarily have a lowest upper bound. Practically, this means that given two hybrid sets f and g, one cannot define their union and their intersection. Fortunately, the sum of hybrid sets is a well-defined operation. Thus, we can easily extend this definition as follows:

Definition 5.1.17 Assume that $A, B : X \to L \times \mathbb{Z}$ are two L-fuzzy hybrid sets. Then their sum, denoted $A \uplus B$, is defined as follows:

$$\bigl(A \uplus B\bigr)(x) = \Bigl(A_\mu(x) \vee B_\mu(x), A_m(x) + B_m(x)\Bigr).$$

Let $\{f_i\}$ denote a finite collection of hybrid sets with a common universe X, where each of these sets contains repeated occurrence of only one element $x_i \in X$. In addition, let us insist that no two f_i and f_j will have common elements. Also, let us denote with $\uplus_i f_i$ the unique hybrid set that is the sum of all f_i. With these preliminary definitions, it is possible to proceed with the following definition:

Definition 5.1.18 Suppose that A is an L-fuzzy hybrid set with universe the set X and that $\alpha \in L$. Then, the α-cut of A, denoted by $^\alpha A$, is the hybrid set $\uplus_i f_i$, where $f_i(x_i) = A_m(x)$ if and only if $\alpha \leq A_\mu(x)$, for all $x_i \in X$.

The α-cut of L-fuzzy hybrid sets has the following properties:

Theorem 5.1.3 *Assume that A and B are two L-fuzzy hybrid sets with universe the set X. Then, the following properties hold:*

(i) if $\alpha \leq \beta$, then $^\alpha A \supseteq {}^\beta A(x)$; and

(ii) $^\alpha(A \uplus B) = {}^\alpha A \uplus {}^\alpha B$.

Proof The proof is similar to the proof of Theorem 5.1.2 and is omitted. □

I will omit the definition of general fuzzy P system, since their definition is just an extension of Definition 5.1.6. The important thing is whether these systems have anything new to offer.

5.2   Fuzzy Labeled Transition Systems

The concept of a fuzzy labeled transition system (FLTS) was independently introduced by Liliana D'Errico and Michele Loreti [45] and this author [127]. The ideas developed below are partially based on ideas and results presented in [92]

5.2.1   Definition and Properties

Let me start with the definition of what an FLTS is.

Definition 5.2.1 A fuzzy labeled transition system over a crisp set of actions $\mathscr{A}$ is a triple $(\mathscr{A}, \mathscr{Q}, \mathscr{T})$ consisting of

- a set $\mathscr{Q}$ of states; and

- a ternary fuzzy relation $\mathscr{T}(\mathscr{Q}, \mathscr{A}, \mathscr{Q})$ called the *fuzzy transition relation*.

If the membership degree of (q, α, q') is d,[2] then $q \xrightarrow[d]{\alpha} q'$ denotes that the plausibility degree to go from state q to state q' by action α is d. More generally, if $q_0 \xrightarrow[d_1]{\alpha_1} q_1 \xrightarrow[d_2]{\alpha_2} q_2 \cdots \xrightarrow[d_n]{\alpha_n} q_n$, then q_n is called a *derivative* of q_0 with plausibility degree equal to $\min\{d_1, d_2, \dots, d_n\}$.

Obviously, one could use any t-norm and, thus, she could provide a more general definition of FLTSes. In addition, one could use a lattice L instead of $[0, 1]$, but the real question is whether these additions would make the resulting structures more expressible or useful.

As in the crisp case, it is quite reasonable to ask when two FLTSes are behaviorally equivalent? In different words, when the behaviors of two distinguishable FLTSes are indistinguishable? Naturally, one is not interested in absolute equivalence but in equivalence to some degree. For example, what can be said about the two FLTSes depicted in Fig. 5.2? In order to be able to answer this question we need to define the notion of similarity between FLTSes.

Definition 5.2.2 Assume that $(\mathscr{A}, \mathscr{Q}, \mathscr{T})$ is an FLTS and that S is a fuzzy binary relation over $\mathscr{Q}$. Then, S is called a *strong fuzzy simulation over* $(\mathscr{A}, \mathscr{Q}, \mathscr{T})$ *with simulation degree s,* denoted as $S_{(s)}$; if $S(p, q) \geq s$ and $p \xrightarrow[d_1]{\alpha} p'$, there exists $q' \in \mathscr{Q}$ such that $q \xrightarrow[d_2]{\alpha} q', d_2 \geq d_1$, and $S(p', q') \geq S(p, q)$. We say that q *strongly fuzzily simulates* p with degree $d \in [0, 1]$, if there exists a strong fuzzy simulation $S_{(s)}$ such that $d \geq s$.

In order to make the notion just defined more clear, I will present an example of a strong fuzzy simulation.

---

2. One could say that the membership degree of a tuple (q, α, q') "indicates the strength of membership within the relation" [75, p. 120].

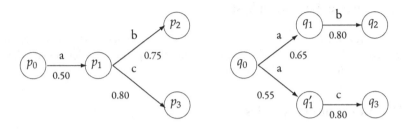

Figure 5.2: Two similar FLTSes

Example 5.2.1 Consider the two FLTSes depicted in Fig. 5.2 and the fuzzy binary relation described by the following table.

S	p_0	p_1	p_2	p_3
q_0	0.4	–	–	–
q_1	–	0.5	–	–
q_1'	–	0.5	–	–
q_2	–	–	0.6	–
q_3	–	–	–	0.6

This fuzzy relation is a strong fuzzy simulation and therefore p_0 strongly fuzzily simulates q_0. To verify this one needs to examine each transition $q_i \xrightarrow{\alpha}_{d} q_j$ for every pair $S(q_i, p_i) > 0$ and show that it is matched by some transition $p_i \xrightarrow{\alpha}_{d'} p_j$. For example, consider the pair (q_1', p_1). State q_1' has one transition $q_1' \xrightarrow{c}_{0.80} q_3$ which is matched by $p_1 \xrightarrow{c}_{0.80} p_3$ because $0.80 \geq 0.80$ and $S(q_3, p_3) \geq S(q_1', p_1)$. Therefore, q_1' strongly fuzzily simulates p_1 with degree 0.80.

Definition 5.2.3 A fuzzy binary relation $S(\mathcal{Q}, \mathcal{Q})$ is said to be a strong fuzzy bisimulation over an FLTS $(\mathcal{A}, \mathcal{Q}, \mathcal{T})$ with simulation degree s if both S and S^{-1} (i.e., the inverse of S) are strong fuzzy simulations. Any p and q are *strongly fuzzily bisimilar with degree d* or *strongly fuzzily equivalent to degree $d \in [0, 1]$, $p \sim_d q$*, if there is a strong fuzzy bisimulation $S_{(s)}$ such that $d \geq s$.

Some strong fuzzy bisimulations are constructed by others more simple ones as the proof of the following propositions shows.

Proposition 5.2.1 *Suppose that each $S_{(s_i)}^{(i)}$, $i = 1, 2, ...$, is a strong fuzzy bisimulation with simulation degree s_i. Then, the following fuzzy relations are all strong fuzzy bisimulations:*

(1) $Id_{\mathcal{Q}}$ (2) $S_{(s)}^{-1}$

(3) $S_{(s_1)}^{(1)} \circ S_{(s_2)}^{(2)}$ (4) $\bigcup_{i \in I} S_{(s_i)}^{(i)}$.

Proof

(i) For the identity relation $\mathrm{Id}_{\mathscr{C}}(\mathscr{C},\mathscr{C})$ it holds that $\mathrm{Id}_{\mathscr{C}}(q,q) = 1$ and $\mathrm{Id}_{\mathscr{C}}(q_i,q_j) = 0$ for all $q, q_i, q_j \in \mathscr{C}$ and where $q_i \neq q_j$. In addition, it holds that $\mathrm{Id}_{\mathscr{C}}^{-1}(\mathscr{C},\mathscr{C}) = \mathrm{Id}_{\mathscr{C}}(\mathscr{C},\mathscr{C})$, which trivially shows that $\mathrm{Id}_{\mathscr{C}}$ is a strong fuzzy bisimulation.

(ii) In order to show that $S_{(s')}^{-1}$ is a strong fuzzy bisimulation we need to show that the inverse of S^{-1} (i.e., S) is a strong fuzzy bisimulation, which is obvious from Definition 5.2.3.

(iii) Recall from Definition 3.4.3 that if $S^{(1)}(\mathscr{P},\mathscr{C})$ and $S^{(2)}(\mathscr{C},\mathscr{R})$ are two strong fuzzy bisimulations with similarity degrees s_1 and s_2, respectively, then $T(\mathscr{P},\mathscr{R}) = S_{(s_1)}^{(1)}(\mathscr{P},\mathscr{C}) \circ S_{(s_2)}^{(2)}(\mathscr{C},\mathscr{R})$ is a new fuzzy binary relation such that

$$T(p,r) = \max_{q \in \mathscr{C}} \min \left[S_{(s_1)}^{(1)}(p,q), S_{(s_2)}^{(2)}(q,r) \right].$$

Obviously, $T(p,r) \geq \min\{s_1, s_2\}$, which shows exactly what we were looking for.

(iv) Assume that $S_{(s_1)}^{(1)}(\mathscr{P},\mathscr{C})$ and $S_{(s_2)}^{(2)}(\mathscr{P},\mathscr{C})$ are two strong fuzzy bisimulations. Then, as expected, their union as fuzzy binary relations is defined as follows:

$$\left(S_{(s_1)}^{(1)} \cup S_{(s_2)}^{(2)} \right)(p,q) = \max \left[S_{(s_1)}^{(1)}(p,q), S_{(s_2)}^{(2)}(p,q) \right].$$

From this it not difficult to see that

$$\left(S_{(s_1)}^{(1)} \cup S_{(s_2)}^{(2)} \right)(p,q) \geq \min \left[S_{(s_1)}^{(1)}(p,q), S_{(s_2)}^{(2)}(p,q) \right]$$

proves the simplest case. From this it is not difficult to see why the general case holds.

$\square$

The following statement reveals some properties of the strong fuzzy bisimulation.

Proposition 5.2.2

(i) $\sim_d$ is an equivalence relation; and

(ii) $\sim_d$ is a strong fuzzy bisimulation.

Proof

(i) Recall that a fuzzy binary relation $R(\mathscr{X},\mathscr{X})$ is a *fuzzy equivalence relation* if it is reflexive, symmetric, and transitive. For reflexivity, it is enough to consider the identity relation $\mathrm{Id}_{\mathscr{C}}(\mathscr{C},\mathscr{C})$, which is a strong fuzzy bisimulation. For symmetry it is enough to say that given a strong fuzzy bisimulation $S_{(s)}$, its inverse $S_{(s')}^{-1}$ is also a strong fuzzy bisimulation. Finally, for transitivity it is enough to say that the relational composition of two strong fuzzy bisimulations is also a strong fuzzy bisimulation.

(ii) This is a direct consequence of Proposition 5.2.1.

$\square$

The notions of fuzzy simulation and a fuzzy bisimulation have been introduced by this author and independently by Miroslav Ćirić, Aleksandar Stamenković, and Jelena Ignjatović, although these authors have used different names and have introduced these notion to describe different things. In particular, the notion of fuzzy simulation is related to the notion of a *right-invariant fuzzy quasi-order* [122] and the notion of a fuzzy bisimulation to the notion of a *right-invariant fuzzy equivalence* [29].

5.2.2   A Fuzzy Process Calculus

Roughly, a process calculus is a formalism to describe concurrent computation. Arthur John Robin Gorell Milner's [91] *calculus of communicating systems*, or just CCS, is an example of such a calculus. D'Errico and Loreti [46] gave a fuzzy version of Milner's CCS. The syntax of this new calculus, which is dubbed FCCS, is given below:

$$A ::= \mathbf{0} \mid X \mid \sum_{i \in I} (\text{act}_i, \lambda_i).A_i$$

$$P ::= Q \mid P_1 | P_2 \mid P \backslash A \mid P[f]$$

$$\text{act} ::= \bar{a} \mid a \mid \tau$$

where

- X is an agent defined by $X \overset{\text{def}}{=} P$; thus, each occurrence of X is replaced by P;

- $\sum_{i \in I} (\text{act}_i, \lambda_i).A_i$ denotes a choice from i possible "behaviors" that start with action act_i with plausibility degree that is equal to λ_i;

- $\mathbf{0}$ is the *inactive agent*, in fact $\mathbf{0} \overset{\text{def}}{=} \sum_{i \in \varnothing} A_i$;

- $P_1 | P_2$ is the *parallel composition* of two processes and denotes the parallel execution of these two processes;

- $P[f]$ is a relabeling operator that can be used to rename process actions;

- $P \backslash A$ is a restriction of P which behaves like P with exception that it is impossible to interact using actions in A and does not include τ; and

- τ is the silent action and is the result of communication of two internal components of some agent.

Milner [91] described the semantics of CCS using *transitional semantics*. In fact, he used labeled transition systems to describe the semantics of his CCS. For each syntactic construct there is a semantic rule that is similar to an inference rule. These semantic rules can be easily adapted to provide an operational semantics of FCCS and such an adaptation is shown in Fig. 5.3.

A few remarks about these semantic rules are in order. First, all rules are in accordance to Kosko's [80] interpretation of fuzzy set theory, that is, the plausibility degree measures the degree to which the particular transition will happen, not whether it will happen or not. Also, the **Act** rule specifies that the plausibility degree of the transition should be less than

$$\textbf{Act} \; \frac{}{(\alpha, \lambda).E \xrightarrow[\lambda']{\alpha} E} \; \lambda' \le \lambda \qquad \textbf{Sum}_j \; \frac{E_j \xrightarrow[\lambda]{\alpha} E'_j}{\sum_{i\in I} E_i \xrightarrow[\lambda]{\alpha} E'i_j} \; j \in I$$

$$\textbf{Com}_1 \; \frac{E \xrightarrow[\lambda]{\alpha} E'}{E|F \xrightarrow[\lambda]{\alpha} E'|F} \qquad \textbf{Com}_2 \; \frac{F \xrightarrow[\lambda]{\alpha} F'}{E|F \xrightarrow[\lambda]{\alpha} E|F'}$$

$$\textbf{Com}_3 \; \frac{E \xrightarrow[\lambda_1]{\alpha} E' \qquad F \xrightarrow[\lambda_2]{\bar{\alpha}} F'}{E|F \xrightarrow[\lambda_1 \wedge \lambda_2]{\tau} E'|F'}$$

$$\textbf{Res} \; \frac{E \xrightarrow[\lambda]{\alpha} E'}{E\backslash L \xrightarrow[\lambda]{\alpha} E'\backslash L} \qquad \textbf{Rel} \; \frac{E \xrightarrow[\lambda]{\alpha} E'}{E[f] \xrightarrow[\lambda]{\alpha} E'[f]}$$

$$\textbf{Con} \; \frac{P \xrightarrow[\lambda]{\alpha} P'}{A \xrightarrow[\lambda]{\alpha} P'} \; (A \overset{\text{def}}{=} P)$$

Figure 5.3: Operational semantics of FCCS

or equal to the degree associated with the prefix. Demanding equality, would make the rule almost nonfuzzy!

There is an extension of CCS, namely the value-passing calculus, where an agent can take *values* as arguments, which can be used to handle different cases. Milner has shown that this feature can be dropped by introducing a *conditional* structure. Providing a satisfactory solution to the problem of value passing in FCCS demands the introduction of the notion of process similarity, that is, a way to see whether two or more processes are similar. This notion will be introduced in Sect. 5.4. Another problem that should be tackled is the introduction of recursion in FCCS. A possible solution to this problem is to replace recursion with replication. Again, how this can be done is discussed in Sect. 5.4.

5.3 Fuzzy X-Machines

The X-machine is a model of computation that has been introduced by Samuel Eilenberg [47]. Roughly, given an arbitrary set X and a family of relations $\Phi = \{\varphi_i\}$ where $\varphi \subseteq X \times X$, an X-machine $\mathcal{M}$ of type Φ is an automaton over the alphabet Φ. Although a labeled transition system is not an automaton (e.g., there are no terminal states), still it is very easy to define automata using the data of a labeled transition system as a starting point. Thus, a fuzzy automaton is a special case of a fuzzy labeled transition system that includes a set of initial states and a set of final states (see [95] for a thorough introduction to fuzzy automata and their theory). More formally,

Definition 5.3.1 A fuzzy automaton is a quintuple $(Q, \Sigma, \delta, q_0, F)$, where

- Q is a finite set of states;

- Σ is a finite set of symbols, called the alphabet of the automaton;

- δ is the fuzzy transition function, that is, $\delta : Q \times \Sigma \times Q \to [0, 1]$;

- $q_0 \in Q$ is the initial state; and

- $F \subseteq Q$ is a set of final states.

Given a fuzzy automaton over an alphabet Σ, a partial function can be defined

$$L_a^{-1}(b) = \left\{ x \mid x \in \Sigma^*, ax = b \right\},$$

where ax is the concatenation of strings a and x. Note that $L_a : \Sigma^* \to \Sigma^*$ is the *left multiplication* and it is defined as $L_a(b) = ab$. Thus, L_a^{-1} is the inverse of the left multiplication. Now, by replacing each edge

$$p \xrightarrow[d]{\alpha} q$$

of an automaton with an edge

$$p \xrightarrow[d]{L_\alpha^{-1}} q,$$

the result is a new automaton which is a fuzzy X-machine. Note that the type of such a machine is $\Phi = \left\{ L_\alpha^{-1} \mid \alpha \in \Sigma \right\}$. Obviously, one can construct an X-machine even from a fuzzy labeled transition system, but the result will not be a *machine* since it will not have initial and terminal states. This view is correct when one has in mind the classical view of a machine as a conceptual device that after processing its input in a *finite* number of operations terminates. Interestingly, there are exceptions to this view that are widely used even today. For example, an operating system does not cease to operate and those that stop unexpectedly are considered failures. Thus, one can assume that a machine will not terminate to deliver a result but instead it delivers (partial) results as it proceeds (see [126, Chap. 5] for more details). On the other hand if states are elements of some fuzzy subset, then we can say that there is a termination degree associated with each computation. In other words, a computation may not completely stop or it may stop at some unexpected moment. This is a novel idea, since the established view is that a computation must either stop or it will loop forever. Ideas like this one could be used to model the case where an external agent abruptly terminates a computation, etc.

5.4   Fuzzy Chemical Abstract Machine

The Gamma model of parallel programming was introduced by Jean-Pierre Benâtre and Daniel Le Métayer [11] (see also [12] for a more accessible account of the model and [13] for a more recent account of it; also see [20] for a thorough presentation of the field of multiset processing). At the time of its introduction, parallel programming, as a mental activity, was considered more difficult (one might also say more cumbersome) than sequential programming. Even today this is still valid to a certain degree. Benâtre and Le Métayer designed Gamma in order to ease the design and specification of parallel algorithms, thus, making a parallel programming task easier compared to previously available approaches. Gamma was inspired by the chemical reaction model. According to this metaphor, the state of a computer system is like a chemical solution in which *molecules*, that is, processes, can interact with each other according to some *reaction rules*, while all possible contacts cam occur since a magic hand stirs the solution continuously. In Gamma solutions are represented by multisets. Processes are elements of a multiset and the reaction rules are multiset rewriting rules. Thus, Gamma is a formalism for multiset processing.

5.4.1   The Chemical Abstract Machine

The *chemical abstract machine*, or *cham* for short, is a model of concurrent computation developed by Gérard Berry and Gérard Boudol [14]. The cham is based on the Gamma model and was designed as a tool for the description of concurrent systems. Basically, each cham is a chemical solution in which floating molecules can interact with each other according to a set of reaction rules. As expected, a magical mechanism stirs the solution so as to allow possible contacts between molecules. Also, as is evident, a number of interactions happen concurrently.

The ingredients of a cham are *molecules* $m_1, \dots, m_n$ that flow in *solutions* $S_1, \dots, S_k$. Solutions change according to a number of *transformation rules*, which determine a *transformation relation* $S_i \rightarrow S_j$. A cham is fully specified by defining its molecules, its solutions, and its transformation rules. There are two types of transformation rules: general rules, applicable to any cham, and local rules, which are used to specify particular chams. Molecules are terms of algebras associated with specific operations for each cham. Solutions are multisets of molecules $[m_k, m_k, \dots, m_l]$. By applying the *airlock* constructor "▷" to some molecule m_l and a solution S_k, one gets a new molecule $m_l \triangleright S_k$.

A specific rule has the following form:

$$m_1, m_2, \dots, m_k \rightarrow m'_1, m'_2, \dots, m'_l,$$

where m_i and m'_i are molecules. Also, there are four general transformation rules:

(i) The *reaction* rule where an instance of the right-hand side of a rule replaces the corresponding instance of its left-hand side. In particular, if there is a rule

$$m_1, m_2, \dots, m_k \rightarrow m'_1, m'_2, \dots, m'_l$$

and if the M_is and M'_js are instances of the m_is and the m'_js by a common substitution, then

$$[M_1, M_2, \dots, M_k] \rightarrow [M'_1, M'_2, \dots, M'_l].$$

(ii) The *chemical* rule specifies that reactions can be performed freely within any solution:

$$\frac{S \to S'}{S \uplus S'' \to S' \uplus S''}.$$

(iii) According to the *membrane* rule sub-solutions can freely evolve in any context:

$$\frac{S \to S'}{[C(S)] \to [C(S')]}'$$

where $C(\ )$ is molecule with a hole $(\ )$ in which another molecule is placed. Note that solutions are treated as megamolecules.

(iv) The *airlock* rule has the following form:

$$[m] \uplus S \leftrightarrow [m \rhd S].$$

5.4.2   On the Similarity of Processes

Informally, a process is a program in execution, which is completely characterized by the value of the *program counter*, the contents of the processor's or core's registers and the process *stack* (see [119] for an overview). Naturally, two or more processes may be initiated from the same program (e.g., think of a web browser or a word processor). In general, such processes are considered as distinct execution sequences. Although they have the same text section, still their data sections will vary. Interestingly, one can view the totality of processes that run in a system at any moment as a multiset that evolves over time, which is built *on-the-fly* and may be viewed as the recorded history of a system. The rationale for this choice is that elements of a multiset may be considered as tokens of different types. Thus, a multiset consisting of two copies of "a" and three copies of "b" can be viewed as a structure that consists of two tokens of type "a" and three tokens of type "b" (see [136] for a thorough discussion of this idea). When dealing with an operating system, one may argue that types represent the various programs available to users and tokens represent the various processes corresponding to these programs.

Although it does make sense to view programs as types and processes as tokens, still not all tokens are identical. For example, different people use a particular web browser to browse different web pages and have different numbers of tabs open at any given moment. Thus, we cannot actually talk about processes that are identical, but we can surely talk about processes that are similar. The next question that needs to be answered is: How can we compare processes? Obviously, the idea of similarity is identical to the idea of fuzzy sets. One can say that there are fuzzy subsets and the various processes belong to a degree to these fuzzy subsets. For example, consider the fuzzy subset of web browsers. Then, each running instance of the browser belongs to this subset to a degree. Nevertheless, the next problem is how one can estimate this degree?

An easy solution is to define an *archetypal* process that "springs" from some binary $\mathcal{b}$ and then to compare each process that "springs" from $\mathcal{b}$ with the archetypal process. The outcome of this procedure will be an estimation of the membership degree to the fuzzy subset of all processes that "spring" from $\mathcal{b}$. But, what exactly is an archetypal process? Clearly, there is no single answer to this question, but roughly one could define it as follows:

Definition 5.4.1 Assume that p is an executable program for some computational platform (operating system, CPU, etc.). Then, an *archetypal process* π of p is the program p running with minimum resources.

Here the term *minimum resources* means that the archetypal process consumes the least possible resources (e.g., memory, CPU time).

Example 5.4.1 Let f be some web browser. Then, an archetypal process of f would be a web browser that starts and shows a blank page in only one tab/window and loads no (external) plug-ins. Similarly, an archetypal process for the Unix command `ls` would be the command running without arguments in an empty directory.

Let $\mathscr{P}_p$ denote the class of all processes initiated from a binary p and let π be the archetypal process of $\mathscr{P}_p$. Then, δ_π is a fuzzy subset of $\mathscr{P}_p$ that measures the degree to which some process p_i is similar to π. In addition, $\Delta_\pi : \mathscr{P}_p \times \mathscr{P}_p \to [0,1]$ is a fuzzy relation that measures the degree to which two processes are similar. For example, one can define Δ_π as follows:

$$\Delta_\pi(p_1, p_2) = 1 - \left| \delta_\pi(p_1) - \delta_\pi(p_2) \right|.$$

Obviously, this definition is not and should not be considered as the final word!

Suppose that one has selected a number of criteria to choose and specify archetypal processes. Also, let $\mathscr{P}_\sigma = \bigcup_{q \in \sigma} \mathscr{P}_q$ denote the set of all possible processes for a particular system σ. Without loss of generality, one can think that the elements of σ are the names of all binaries that can possibly be executed on the system. For instance, for some typical Unix system, the class σ may contain the names of programs under /usr/bin, /usr/local/bin, /opt/sfw/bin, etc. Suppose that $\pi \in \sigma$ and that at some moment t, $p_1, \ldots, p_n$ are processes initiated from program π. Then, this situation can naturally be modeled by a *fuzzy multiset* [143], that is, a structure that is characterized by a higher-order function $z : \mathscr{P}_\sigma \to ([0,1] \to \mathbb{N})$. By uncurrying the functional z, one gets a function $\zeta : \mathscr{P}_\sigma \times [0,1] \to \mathbb{N}$. Thus, in general, at any given moment the processes running in a system can be described by a multiset of pairs (p_i, ℓ_i), where ℓ_i denotes the membership degree. However, such a structure reflects what is happening in a system at a particular moment. Thus, to describe what is going on in a system at some time interval, one has to use a structure that can reflect changes as time passes. The most natural choice that can solve this problem is a form of a *set through time*. The paragraph that follows assumes familiarity with topology and category theory. Readers not familiar with these subjects can safely skip the next paragraph.

Bill Lawvere was the first to suggest that sheaves can be viewed as continuously varying sets (see [5, 10] for a detailed account of this idea). Since in this particular case I am interested in *fuzzy multisets continuously varying through time*, it seems that sheaves are not the structures I am looking for. However, as I will show this is not true. But before I proceed, it is more than necessary to give the general definition of a sheaf. The definition that follows has been borrowed from [55] (readers not familiar with basic topological notions should consult any introductory textbook; e.g., see [88]):

Definition 5.4.2 Let X be a topological space and $\mathscr{O}(X)$ its collection of open sets. A *sheaf* over X is a pair (T, p) where T is a topological space and $p : T \to X$ is a local homeomorphism (i.e., each $t \in T$ has an open neighborhood U in T that is mapped homeomorphically by p onto $p(U) = \{p(x) \mid x \in U\}$, and the latter is open in X).

Now it is possible to construct a fuzzy multiset through time. Suppose that $A : X \to [0,1] \to \mathbb{N}$ characterizes some fuzzy multiset. Clearly, function $A' : X \times [0,1] \to \mathbb{N}$ also characterizes the same fuzzy multiset and so does the graph of this function. Let M_j, $j \in J$, where J is a set of indices, be the graphs of all functions A'_j that characterize fuzzy multisets. In addition, assume that each M_j is an open set of a topological space $\mathscr{X}$. Obviously, it is not difficult to define such a topological space. For example, it is possible to define a metric between points $((x_k, i_k), n_k)$ and $((x_l, i_l), n_l)$ and from this to define a metric topology. Having defined a topology on $(X \times [0,1]) \times \mathbb{N}$, it is straightforward to define a sheaf over $\mathscr{X}$. In particular, if $\mathbb{N}$ denotes the order topology on the set of natural numbers, then $\mathscr{N} = (\mathbb{N}, p)$, where $p : \mathbb{N} \to \mathscr{X}$ is a local homeomorphism, is a sheaf over $\mathscr{X}$. In general, such a sheaf characterizes a fuzzy multiset through *discrete* time. Clearly, this is not the only sheaf over $\mathscr{X}$ one can define. In fact, one can build a category $\mathbf{Sh}(\mathscr{X})$ with objects all sheaves over $\mathscr{X}$ and with arrows $k : (A, p) \to (B, q)$ the continuous maps $k : A \to B$ such that

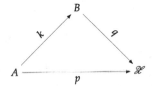

commutes. In general, the sheaf $\mathscr{R} = (\mathbb{R}_0^+, q)$, where $\mathbb{R}_0^+$ is the set of the positive real numbers including zero which also denotes the order topology on this set and $q : \mathbb{R}_0^+ \to \mathscr{X}$ is a local homeomorphism, characterizes a fuzzy multiset through *continuous* time. It is not difficult to define a monic arrow $k : \mathscr{N} \to \mathscr{R}$ and, thus, to show that k belongs to $\mathrm{Sub}(\mathscr{R})$, that is, the collection of sub-objects of $\mathscr{R}$. Thus, we have an implicit proof of the following statement:

Corollary 5.4.1 *Sheaf $\mathscr{R}$ contains more information than $\mathscr{N}$.*

To put it differently, no discrete system can fully simulate a continuous system. In different words, if time and space are continuous (see [126, Sect. 8.3] for a discussion of the discrete vs. continuous controversy in physics), this implies that no real system can be implemented in a digital computer system. Also, it should be obvious that in the most general case one cannot know beforehand all the components of a sheaf representing a fuzzy multiset through time. When such a structure is used to represent processes, then it is *noncomputable*, since one cannot construct it using some algorithm as it is not possible to foresee what the users will do, unless we live in a completely deterministic universe...

5.4.3   The Fuzzy Chemical Abstract Machine

Roughly, a fuzzy cham, or just *fucham*, can be identified with a solution of fuzzy molecules and a set of fuzzy reaction rules. A solution of fuzzy molecules can be modeled by a fuzzy multiset, while fuzzy reaction rules can be described by fuzzy transitions. Before presenting a formal definition of fuchams let us informally examine whether it makes sense to talk about solutions of fuzzy molecules and a plausibility for each reaction rule. To begin with consider the following concrete chemical reaction rule:

$$2H_2 + O_2 \rightarrow 2H_2O.$$

According to the "traditional" view, two hydrogen molecules react with one oxygen molecule to create two water molecules. A fuzzy version of this reaction rule should involve fuzzy molecules and it should be associated them with plausibility degrees. This is justified by the *fact* that molecules of some chemical element or compound are not identical but rather similar to a degree with an ideal molecule of this particular element or compound. In other words, not all hydrogen and oxygen molecules that react to create water are identical. For example, think of deuterium and tritium as "hydrogen" molecules up to some degree that react with oxygen to produce heavy water, tritiated water, and/or partially tritiated water, that is, water up to some degree. Thus, the "water" molecules produced when millions of "hydrogen" molecules react with oxygen molecules are not identical but just similar (if, in addition, the reader considers hydrogen peroxide, that is, H_2O_2, then things will get really *fuzzy*). Obviously, the higher the similarity degree, the more likely it is that the reaction will take place. And this is the reason one must associate with each reaction rule a plausibility degree. Although these ideas may seem unnatural, still Gianfranco Cerofolini and Paolo Amato [22] have used fuzziness and *linear logic*[3] to propose an axiomatic theory for general chemistry. In particular, they have developed ideas similar to mine in order to explain how chemical reaction takes place, which means that my proposal, which I call the *fuzzy chemical metaphor*, is not unnatural at all.

The fuzzy chemical metaphor is essentially identical to the chemical metaphor; nevertheless, it assumes that molecules of the same kind are similar and not identical. Solutions of fuzzy molecules may react according to a number of fuzzy reaction rules, whereas each rule is associated with a feasibility degree that specifies how plausible it is that a particular reaction will take place. A fucham is an extension of the (crisp) cham that is build around the fuzzy chemical metaphor. Like its crisp counterpart, any fucham may have up to four different kinds of transformation rules that are described below.

Fuzzy Reaction Rules   Assume that we have a solution with fuzzy molecules that are supposed to react according to some fuzzy reaction rule. Then, the reaction will take place only when the similarity degree of each molecule is greater or equal to the feasibility of the particular reaction rule.

Definition 5.4.3   Let $(m_i)_{i=1,...,k}$ and $(m'_j)_{j=1,...,l}$ be archetypal molecules. Then,

$$m_1, \ldots, m_k \underset{\lambda}{\rightarrow} m'_1, \ldots, m'_l,$$

is an *ideal* fuzzy reaction rule with feasibility degree λ that describes how likely it is that molecules (m_i) may react to create molecules (m'_j). Suppose that M_i is an instance of m_i to

---

3. "Linear logic appeared as a by-product of coherent semantics. The novelty was the emphasis on structural rules, thus individuating linear negation. Linear logic is spiritual, like classical and intuitionistic logics" [52, p. 442].

degree $\delta_{m_i}(M_i)$, that is, the molecule M_i is similar to m_i with degree equal to $\delta_{m_i}(M_i)$. Then, the following fuzzy reaction

$$[M_1, \dots, M_k] \underset{\lambda}{\to} [M_1', \dots, M_l']$$

is *feasible* with plausibility degree equal to λ if

$$\min\{\delta_{m_1}(M_1), \dots, \delta_{m_k}(M_k)\} \geq \lambda. \tag{5.1}$$

The similarity degree of a molecule M_j' depends on the similarity degrees of the atoms that make up this particular molecule.

Remark 5.4.1 Obviously, in the previous definition, one can replace min with a t-norm.

It is quite possible to have a situation where the same reacting molecules may be able to yield different molecules, something that may depend on certain factors. In different words, we may have a solution where a number of different fuzzy reaction rules are *potentially applicable*. In this case, the reaction rule with the highest plausibility degree is *really applicable*.

Definition 5.4.4 Assume that S is a solution for which the following reaction rules are potentially applicable

$$m_1, \dots, m_k \underset{\lambda_1}{\to} m_1', \dots, m_{l_1}'$$

$$m_1, \dots, m_k \underset{\lambda_2}{\to} m_2'', \dots, m_{l_2}''$$

$$\vdots$$

$$m_1, \dots, m_k \underset{\lambda_n}{\to} m_1^{(n)}, \dots, m_{l_n}^{(n)}$$

and that $\delta_{m_i}(M_i)$, $i = 1, \dots, k$ are the similarity degrees of the actual molecules that are contained in S. Then the really applicable rule is the one that satisfies the conditions of definition (5.1) and whose feasibility degree is the largest among the feasibility degrees of all potentially applicable rules.

Using the Perl pseudo-code of Algorithm 5.4.1, one can compute the really applicable and the potentially applicable rules.

The Fuzzy Chemical Rule   Mixing up two solutions S_1 and S_2 yields a new solution S_3 that contains the molecules of both S_1 and S_2. In other words, S_3 is the sum of S_1 and S_2, or more formally

$$S_3 = S_1 \uplus S_2.$$

Note that in order to find the sum of two or more fuzzy multisets we work as in the crisp case. Nonetheless, because of restriction (5.1), the fuzzy chemical rule takes the following form:

$$\frac{\left(S_1 \underset{\lambda}{\to} S_2\right) \quad \left(\forall M_i \in S_3 : \delta_{m_i}(M_i) \geq \lambda\right)}{S_1 \uplus S_3 \underset{\lambda}{\to} S_2 \uplus S_3}.$$

```
$ξ = min{M_1, ... , M_k} # similarity degrees
$i = 1
$R = -1
@Rules = ()
$Λ = 0
foreach $λ (λ_1, ... , λ_n) {# feasibility degrees
    if (ξ ≥ λ) {
        push @Rules, [$λ, $i]
        if ($λ ≥ $Λ) {
            $Λ = $λ;    $R = $i
        }
    }
    $i + +
}
```

Algorithm 5.4.1: A Perl pseudo-code that computes the really applicable rule of set of rules

Note that this restriction applies to all other general rules.

The Fuzzy Airlock Rule   The airlock rule creates a new molecule that consists of a solution and a molecule. Therefore, one needs to define a similarity degree for solutions in order to be able to estimate the similarity degree of the new molecule.

Definition 5.4.5 Suppose that P is a process solution represented by a fuzzy multiset S. Then, the similarity degree of P to a solution that contains only prototype molecules is given by

$$\Delta_\delta(P) = \min\left\{\delta_{m_i}(M_i) \mid M_i \in P \text{ and } i = 1, 2, ...\right\}.$$

If S is a fuzzy solution and m a fuzzy molecule, then

$$\frac{\lambda \leq \min\left\{\Delta_\delta(S), \delta_m(M)\right\}}{[M] \uplus S \underset{\lambda}{\leftrightarrow} [M \rhd S]}.$$

The Fuzzy Membrane Rule   Suppose that $\delta_C(C())$ denotes the similarity degree of molecule $C()$. Then, the fuzzy membrane rule is formulated as follows:

$$\frac{\left(S \underset{\lambda}{\rightarrow} S'\right) \quad \left(\lambda \leq \min\left\{\Delta_\delta(S), \delta_\pi(C())\right\}\right)}{[C(S)] \underset{\lambda}{\rightarrow} [C(S')]}.$$

5.4.4　Fuzzy π-Calculus

One of the immediate applications of the cham was its use as a tool for the semantic description of various process calculi. In particular, the cham has been used to give an operational semantics of TCCS, the π-calculus, and the concurrent λ-calculus (see [14]). Since the cham is a special case of the fucham, it does make sense to see the use of the fucham as a tool for the semantic description of fuzzy process calculi. Although I have presented a fuzzy process calculus in Sect. 5.2.2, still it does make sense to see how one could build the equivalent of the π-calculus in a fuzzy setup. In addition, it is quite interesting to see how one can give a semantic description of this new calculus in terms of the fucham.

The π-calculus is a mathematical formalism especially designed for the description of *mobile processes*, that is, processes that live in a virtual space of linked processes with mobile links. The π-calculus is a basic model of computation that rests upon the primitive notion of *interaction*. It has been argued that interaction is more fundamental than reading and writing a storage medium; thus, the π-calculus is more fundamental than Turing machines and the λ-calculus (see [126] and the references herein for more details).

In the π-calculus processes are described by process expressions that are defined by the following abstract syntax:

$$P ::= \sum_{i \in I} \pi_i.P_i \,\Big|\, P_1 \,|\, P_2 \,\Big|\, \text{new } \alpha \, P \,\Big|\, !P.$$

If $I = \varnothing$, then $\sum_{i \in I} \pi_i.P_i = 0$, where 0 is the null process that does nothing. In addition, π_i denotes an *action prefix* that represents either sending or receiving a message, or making a silent transition:

$$\pi \quad ::= \quad \begin{array}{ll} x(y) & \text{receive } y \text{ along } x \\ \bar{x}\langle y \rangle & \text{send } y \text{ along } x \\ \tau & \text{unobservable action.} \end{array}$$

The expression $\sum_{i \in I} \pi_i.P_i$ behaves just like one of the P_i's, depending on what messages are communicated to the composite process; the expression $P_1 \,|\, P_2$ denotes that both processes are concurrently active; the expression new α P means that the use of the message α is restricted to the process P; and the expression $!P$ means that there are infinitely many concurrently active copies of P. This is exactly one reason why the π-calculus is strictly more powerful than the λ-calculus. Derivations in the π-calculus follow a very small set of reaction rules, which are given in the table that follows.

Figure 5.4: Different forms of the same character drawn from different fonts that demonstrate the notion of typographic similarity

$$\begin{array}{c} \text{TAU: } \tau.P + M \to P \\ \text{REACT: } \big(x(y).P + M\big) \mid \big(\bar{x}\langle z\rangle.Q + N\big) \to \{z/y\}P|Q \\ \text{PAR: } \dfrac{P \to P'}{P\,|\,Q \to P'\,|\,Q} \qquad \text{RES: } \dfrac{P \to P'}{\text{new } x\, P \to \text{new } x\, P'} \\ \text{STRUCT: } \dfrac{P \to P'}{Q \to Q'} \text{ if } P \equiv Q \text{ and } P' \equiv Q'. \end{array}$$

The expression $P \equiv Q$ means that the process expressions P and Q are structurally congruent. This, in turn, means that one can transform one expression into the other by using the following rules in any direction:

(i) change of bound names (α-conversion);

(ii) $P\,|\,0 \equiv P, P\,|\,Q \equiv Q\,|\,P, P\,|\,(Q\,|\,R) \equiv (P\,|\,Q)\,|\,R$;

(iii) new $x\,0 \equiv 0$, new $x\,y\,P \equiv$ new $y\,x\,P$,[4] new $x\,(P\,|\,Q) \equiv P\,|$ new $x\,Q$ if x is not free in P; and

(iv) $!P \equiv P\,|\,!P$.

As it stands the only way to introduce fuzziness in the π-calculus is to assume that action prefixes are fuzzy. Usually, it is assumed that there is an infinite set of names $\mathcal{N}$ from which the various names are drawn. In our case, it can be assumed that names are drawn from $\mathcal{N} \times [0, 1]$. In other words, a name would be a pair (x, i) which will denote that the name used is similar to the prototype x with degree equal to i. Skeptic readers may find this idea strange as an x is always an x and nothing more or less. Indeed, this is true; nevertheless, if we consider various xs drawn from different (computer) fonts, then one x is more x than some other x. To fully understand this idea, consider the sequence of letters in Fig. 5.4 borrowed from [126]. Clearly, the rightmost symbol does not look like an "A," at least to the degree the second and the third from the left look like an "A." I call this kind of similarity *typographic similarity*, for obvious reasons. Thus, one can say that names are typographically similar.

Berry and Boudol [14] provide two encodings of the π-calculus as a cham, but for reasons of brevity I will consider only one of them. The following rules can be used to describe the functionality of the π-calculus.

$p\,|\,q \rightleftharpoons p, q$	(parallel)
$x(y).p, \bar{x}\langle z\rangle.q \to p[z/y], q$	(reaction)
$!p \rightleftharpoons p, !p$	(replication)
$0 \rightharpoonup$	(inaction cleanup.)
new $x\,p \rightleftharpoons$ new $y\,p[y/x]$ if y is not free in p	(α-conversion)
new $x\,p \rightleftharpoons$ new $x\,[p]$	(restriction membrane)
new $x\,S, p \rightleftharpoons$ new $x\,[p \rhd S]$ if x is not free in p	(scope extension)

---

4. Note that new $x\,y\,P$ is a shorthand for new x new $y\,P$.

The first four rules are shared by the two encodings of Berry and Boudol. Unfortunately, these rules do not describe summation. For example, one can imagine that a sum is an inactive megamolecule that changes to a simpler molecule in a single step.

This encoding could be used to encode a fuzzy version of the π-calculus (i.e., the one obtained by replacing each name with a pair consisting of a name and a similarity degree). Naturally, one can use the rules presented above as they are, but it seems that somehow fuzziness will be lost by adopting this encoding.

Basically, the reaction and α-conversion rules are the most problematic rules. A fuzzification of these rules can be obtained by attaching to each rule a plausibility degree. In the first case, it is reasonable to demand that the similarity degrees of x and $\bar{x}$ are the same and at the same time greater than the plausibility degree and also the difference of the similarity degrees of y and z is not greater than the plausibility degree. In other words, the reaction

$$(x, i_x)\Big((y, i_y)\Big).p, (\bar{x}, i_{\bar{x}})\langle(z, i_z)\rangle.q \xrightarrow{\lambda} p[(z, i_z)/(y, i_y)], q$$

is feasible if $i_x = i_{\bar{x}} \geq \lambda$ and $i_z - i_y \leq \lambda$. Similarly, the α-conversion

$$\text{new}(x, i_x)p \underset{\lambda}{\rightleftharpoons} \text{new}(y, i_y)p[(y, i_y)/(x, i_x)]$$

is plausible only if $i_y - i_x \leq \lambda$. Because of this definition, it is necessary to define the notion of fuzzy structural congruence. One option is to use a slightly modified version of the definition provided in [92]. The slight modification involves α-conversion and the "equation" new x new y $P \equiv$ new y new x P. This implies that

$$\text{new}(x, i_x)\,\text{new}(y, i_y)\,P \equiv_\lambda \text{new}(y, i_y)\,\text{new}(x, i_x)\,P$$

if and only if $\min\{i_x, i_y\} \geq \lambda$.

A. Computing with Words

Generally speaking, computing with words is problem-solving methodology that is supposed to help solving problems where fuzzy sets and linguistic expressions and not numbers best describe them.

A.1   What Is Computation?

It is quite possible that the discussion in Sect. 1.1 has given to the reader the impression that a computation is a mechanical procedure that processes numbers. The truth is that computation is more about symbol manipulation than number processing. After all, Hilbert was dreaming of a fully mechanized mathematical science, which simply means the mechanical manipulation of symbols and not just numbers. Although there are various definitions of what computation is, still I find the following definition of computation by Stevan Harnad's very precise and intuitive:

Definition A.1.1 Computation is an implementation-independent, systematically interpretable, symbol manipulation process [64].

Zadeh introduced the concept of *linguistic variable* (e.g., see [146]) as a means to measure "quantities" such as age when the possible values are expressed in words, for example, young, not young, very young, quite young, old, and not very old. Based on the remark that the human brain mostly manipulates *perceptions* (e.g., perceptions of size, smell, color) and not numbers and measurements, he proposed his *computing with words* [149, 150, 152] methodology as means to solve specific problems in a natural way. Broadly speaking, Zadeh's goal is to make the computing with words methodology a *computational theory of perceptions*. But what does computing with words mean?

A.2   What Is Computing with Words?

According to Zadel [150] "humans employ mostly words in computing and reasoning, arriving at conclusions expressed as words from premises expressed in natural language or having the form of mental perceptions." Nevertheless, something very crucial is missing from this remark—in order to perform some computation, it is necessary to perform some sort of deduction, but it is not known how exactly humans do reason. According to Byrne and Johnson-Laird [19] there are three different theories that (try to) explain how people reason. In particular, *formal rule* theories are based on the assumption that the mind uses a set of formal inference rules, similar to those employed in formal logic. The rules are applied when

A. Syropoulos, *Theory of Fuzzy Computation*, IFSR International Series on Systems Science and Engineering 31, DOI 10.1007/978-1-4614-8379-3,
© Springer Science+Business Media New York 2014

the mind finds that a proposition matches them. The theory of *mental models* depends on conceiving an image or a picture of, especially as a future, possibility. Given some premises, an individual constructs a model of the possibilities to which these premises refer and she draws conclusions that make sense in these possibilities. A person rejects a conclusion when she can find a counterexample, that is, a possibility in which the premise holds but the conclusion does not. *Suppositional* theories suggest that conditionals bring out suppositional thinking. For instance, one thinks that some if-clause is true and, therefore, thinks about the consequences. One such consequence might be that one reaches a contradiction, which *implies* that she has to reject the corresponding supposition. Byrne and Johnson-Laird [19] have argued that the theory of mental models is our best chance to resolve the puzzle of "if."

Since it is not known for sure how intelligent beings reason, one cannot hope to build machines and systems that will reason as well as humans. Thus, even if one "computes" with words, numbers, or icons(!), it is not sure whether she will be able to perform the required actions. Indeed, Zadeh [150] correctly remarks that "we cannot automate driving in heavy traffic; we cannot translate from one language to another at the level of a human interpreter; we cannot create programs which can summarize non-trivial stories; our ability to model the behavior economic systems leaves much to be desired; and we cannot build machines that can compete with children in the performance of a wide variety of physical and cognitive tasks." Whereas he argues that we need fuzziness to solve, or at least tackle, these problems, I think that we need to first fully understand a number of *natural phenomena* that are obviously vague, in order to solve these problems. In different words, I do not think that fuzziness alone is enough to handle these problems.

The concept of a *granule* plays a crucial importance in computing with words. Generally speaking, a granule is a cluster of objects, or points if you prefer this term, that are similar, indistinguishable, close enough, etc. For example, the terms young, old, very old, middle aged, etc., may form a granule of age. Formally, a granule is defined as follows:

Definition A.2.1 A granule is a clump of objects defined by a generalized constraint

$$x \text{ irs } R,$$

where x is a constrained variable, R is a constraining relation, and "irs" is a copula.

In computing with words, granules are viewed as fuzzy constraints on variables. For example, in the sentence "Mary is young," which can be viewed as a linguistic characterization of a perception, the word *young* is the label of a granule called *young*. The fuzzy set *young* can be viewed as a fuzzy constraint on the age of Mary.

The problems that can be tackled with computing with words consists of a set of propositions that are expressed in natural language (e.g., Swedish), which is called the *initial data set*. Using this data set, a program implementing this programming paradigm should be able to answer in natural language a query that is also expressed in natural language.

Although computing with words is not a model of computation, it is supposed to be a programming paradigm. However, this paradigm is typically implemented on crisp hardware using crisp software, while all conceptual machines presented so far were vague by definition. And this is exactly the reason why computing with words is problematic. If it were to run on fuzzy hardware as fuzzy software, whatever this might mean, then it would definitely make sense.

B. The Rough Sets Approach

Rough sets are an alternative mathematical formulation of vagueness in which a vague entity (e.g., a set) is described by two crisp sets: the *upper* and *lower* approximations. Although rough sets were introduced by Zdzisław Pawlak [103] in 1982, only recently Sumita Basu [7] has described automata based on rough sets. In this appendix, I will briefly present rough sets, automata based on rough set theory, and a first attempt at defining rough Turing machines.

B.1 Rough Sets

Assume that X is a universe and R is an equivalence relation on X (see Footnote 1 on p. 50). An equivalence relation R partitions X into blocks of R-related elements, called equivalence classes. Typically, the equivalence class of an element a is denoted $[a]$ and is defined as the set

$$[a] = \{x \mid (x \in X) \wedge (a \, R \, x)\}.$$

The pair (X, R) is called an *approximation space* and R is called an *indiscernibility* relation. When $x, y \in X$ and $x \, R \, y$, x and y are indistinguishable in (X, R). The set of all equivalence classes X, denoted X/R and called a *quotient* set, forms a partition of X. Any element $E_i \in X/R$ is called an *elementary set*.

If $A \subseteq X$, the *upper approximation*, $\overline{A}$, and the *lower approximation*, $\underline{A}$, of A are defined as follows:

$$\overline{A} = \bigcup_{E_i \cap A \neq \varnothing} E_i = \{x \mid (x \in X) \wedge ([x]_R \cap A \neq \varnothing)\},$$

$$\underline{A} = \bigcup_{E_i \subseteq A} E_i = \{x \mid (x \in X) \wedge ([x]_R \subseteq A)\}.$$

That is, the upper approximation of A is the union of R-equivalence classes whose intersection with A is nonempty while the lower approximation of A is the union of R-equivalence classes contained in A. An element $x \in X$ definitely belongs to A when $x \in \underline{A}$. The same element possibly belongs to A when $x \in \overline{A}$. The *boundary* Bnd(A) of A is the set $\overline{A} \setminus \underline{A}$. A set $A \subseteq X$ is said to be *definable* in (X, R) if and only if Bnd(A) = $\varnothing$.

Definition B.1.1 Given an approximation space (X, R), a rough set in it is a pair (L, U) such that L and U are definable sets, $L \subseteq U$, and if any R-equivalence class is a singleton $\{s\}$ and $\{s\} \in U$, then $\{s\} \in L$.

A. Syropoulos, *Theory of Fuzzy Computation*, IFSR International Series on Systems Science and Engineering 31, DOI 10.1007/978-1-4614-8379-3,
© Springer Science+Business Media New York 2014

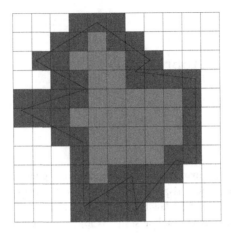

Figure B.1: A graphical representation of a rough set. The *solid line* shows the boundary of a set A, while "crisp" sets are defined with squares. From the *colored squares*, those that are painted with *lighter color* form the lower approximation of A and those that are painted with the *darker color* form the difference between the upper and the lower approximations

Alternatively, one could say that these two sets define the rough set as they specify its lower and upper boundaries. Figure B.1 depicts the boundary of a set A, its lower approximation, and, essentially, its upper approximation. The *accuracy of approximation* [104]

$$\alpha_R(A) = \frac{\left|\underline{A}\right|}{\left|\overline{A}\right|}$$

characterizes the accuracy of the representation. Clearly, the number $\alpha_R(A)$ is greater or equal to zero and less than or equal to one. In the particular case where $\alpha_R(A) = 1$, set A is crisp with respect to R.

The membership function for rough sets is defined by employing the relation R as follows:

$$\mu_A^R(x) = \frac{\left|A \cap [x]\right|}{\left|[x]\right|}.$$

It is obvious that $0 \leq \mu_A^R(x) \leq 1$.

Didier Dubois and Henri Prade [43] have presented the idea that fuzzy sets and rough sets "aim to different purposes." Thus, they have introduced what they call *rough fuzzy* and *fuzzy rough* sets. However, it is pretty obvious that both fuzzy and rough sets aim to describe vagueness and combining the two aims to describe what? I think that both approaches to vagueness have their merits but combining them yields just a mathematical generalization and nothing more. On the other hand, it would make sense, for example, to define rough multisets as this would correspond to something one can directly observe. Thus, one could use them to define rough P systems, etc.

B.2   Rough Finite-State Automata

Let us start with the definition of rough finite-state semi-automata:

Definition B.2.1 The quadruple $A = (Q, R, S, \delta)$ is a *rough finite-state semi-automaton* if Q is a finite set of internal states, R is an equivalence relation on Q, S is the set of input symbols, **D** is the class of all definable sets in the approximation space (Q, R), and $\delta : Q \times S \to \mathbf{D} \times \mathbf{D}$ is the transition function, and it holds that for all $q \in Q$ and all $\sigma \in S$, $\delta(q, \sigma) = (D_1, D_2)$, where $D_1, D_2 \in \mathbf{D}$ and $D_1 \subseteq D_2$.

Obviously, $\delta(q, \sigma)$ is a rough set having as lower and upper approximations the sets D_1 and D_2, respectively, that is,

$$D_1 = \underline{\delta(q, \sigma)} \quad \text{and} \quad D_2 = \overline{\delta(q, \sigma)}.$$

Also, the transition function of this automaton yields a rough set of states (compare this function with the transition function of a fuzzy automaton; see Definition 5.3.1 on p. 124). Moreover, if for all $q \in Q$ and $\sigma \in S$

$$\delta(q, \sigma) = ([q'], [q']),$$

where $q' \in Q$ and $[q'] = \{q'\}$, then this automaton is actually a finite-state automaton.

Definition B.2.2 The extended transition function $\delta^* : \mathbf{D} \times S \to \mathbf{D} \times \mathbf{D}$, for some equivalence class $[a]$ of R and $\sigma \in S$, is defined as follows:

$$\delta^*([a], \sigma) = \left(\underline{\delta^*([a], \sigma)}, \overline{\delta^*([a], \sigma)} \right),$$

where

$$\underline{\delta^*([a], \sigma)} = \bigcup_{\alpha \in [a]} \underline{\delta(\alpha, \sigma)}$$

and

$$\overline{\delta^*([a], \sigma)} = \bigcup_{\alpha \in [a]} \overline{\delta(\alpha, \sigma)}.$$

Also, for any definable set D

$$\underline{\delta^*(D, \sigma)} = \bigcup_{[a] \subseteq D} \underline{\delta^*([a], \sigma)}$$

and

$$\overline{\delta^*(D, \sigma)} = \bigcup_{[a] \subseteq D} \overline{\delta^*([a], \sigma)}.$$

It is possible to extend the transition function from $Q \times S \to \mathbf{D} \times \mathbf{D}$ to $Q \times S^* \to \mathbf{D} \times \mathbf{D}$ as follows:

(i) for all $q \in Q$, $\delta(q, \varepsilon) = ([q], [q])$ and for any $x \in S^*$, $\delta(q, x) = \delta([q], x)$; and

(ii) for all $q \in Q$, all $\sigma \in S$ and all $x \in S^*$

$$\delta(q, x\sigma) = \left(\underline{\delta(q, x\sigma)}, \overline{\delta(q, x\sigma)} \right),$$

where

$$\underline{\delta(q, x\sigma)} = \bigcup_{B \subseteq \delta(q,x)} \underline{\delta(B, \sigma)} = \underline{\delta(\delta(q, x), \sigma)}$$

and

$$\overline{\delta(q, x\sigma)} = \bigcup_{B \subseteq \delta(q,x)} \overline{\delta(B, \sigma)} = \overline{\delta(\overline{\delta(q, x)}, \sigma)}.$$

Given a rough semi-automaton, an important question is which states are reachable from a particular state? I think that one should expect that there are two sets—the set of states that are definitely reachable and the set of states that are possibly reachable.

Definition B.2.3 Assume that $A = (Q, R, S, \delta)$ is a rough semi-automaton. Then, a state $q' \in Q$ is a *definite successor* of $q \in Q$ if there is a $x \in S^*$ such that $q' \in \underline{\delta(q, x)}$ or $[q'] \subseteq \underline{\delta^*([q], x)}$. The set of all definite successors of $q \in Q$ is denoted by $D(q)$. For $Q' \subseteq Q$,

$$D(Q') = \bigcup_{q \in Q'} D(q).$$

It is easy to see that when q_1 is a definite successor of q and q_2 is a definite successor of q_1, then q_2 is a definite successor of q.

Definition B.2.4 Assume that $A = (Q, R, S, \delta)$ is a rough semi-automaton. Then, a state $q' \in Q$ is a *possible successor* of $q \in Q$ if there is a $x \in S^*$ such that $q' \in \overline{\delta(q, x)}$ or $[q'] \subseteq \overline{\delta^*([q], x)}$. The set of all possible successors of $q \in Q$ is denoted by $P(q)$. For $Q' \subseteq Q$,

$$P(Q') = \bigcup_{q \in Q'} P(q).$$

As in the case of definite successors, when q_1 is a possible successor of q and q_2 is a possible successor of q_1, then q_2 is a possible successor of q. In addition, when q_1 is a definite successor of q, then q_1 is a possible successor of q, but not vice versa.

By adding a set of accepting or halting states and a set of initial states to a rough semi-automaton, the resulting structure becomes a rough automaton.

Definition B.2.5 The sextuple $A = (Q, R, S, \delta, I, H)$ is a *rough finite-state automaton* if the quadruple (Q, R, S, δ) is a rough finite-state semi-automaton, I is a definable set in the approximation space (Q, R) that is called the *initial configuration*, and $H \subseteq Q$ is the set of *accepting states* of A.

For the rest of this section it is necessary to define a binary relation T on S^* as follows:

$$x \, T \, y \quad \text{iff} \quad \underline{\delta^*(I, x)} = \underline{\delta^*(I, y)} \text{ and } \overline{\delta^*(I, x)} = \overline{\delta^*(I, y)}.$$

This relation is an equivalence relation on S^*.

Definition B.2.6 The behavior of a rough automaton A, β_A, is defined by

$$\beta_A = (\underline{\beta_A}, \overline{\beta_A})$$

where

$$\underline{\beta_A} = \left\{ x \mid (x \in S^*) \wedge \left(\underline{\delta^*(I, x)} \cap H \neq \varnothing \right) \right\},$$

$$\overline{\beta_A} = \left\{ x \mid (x \in S^*) \wedge \left(\overline{\delta^*(I, x)} \cap H \neq \varnothing \right) \right\}.$$

Theorem B.2.1 *Given a rough automaton A, its behavior β_A is a rough set on (S^*, T).*

Proof It is enough to show that for some $x \in S^*$, $x \in \underline{\beta_A}$ implies that $[x]_T \subseteq \underline{\beta_A}$ and $x \in \overline{\beta_A}$ implies that $[x]_T \subseteq \overline{\beta_A}$. This means that both $\underline{\beta_A}$ and $\overline{\beta_A}$ are unions of equivalence classes of T. Suppose that $x \in \underline{\beta_A}$. Then, this means that $\underline{\delta^*(I, x)} \cap H \neq \varnothing$, and if $y \in [x]_T$ of $x \, T \, y$, then $\underline{\delta^*(I, x)} = \underline{\delta^*(I, y)}$ and, thus, $\underline{\delta^*(I, y)} \cap H \neq \varnothing$, which implies that $y \in \underline{\beta_A}$ and so $[x]_T \subseteq \underline{\beta_A}$, which proves the first part. The second part can be proved in a similar way. $\quad\square$

Definition B.2.7 Assume that $A = (Q, R, S, \delta, I, H)$ is a rough finite-state machine and that R' is an another equivalence relation on Q such that $R' \supseteq R$ or $R' \subseteq R$. Then, $A/R' = (Q, R', S, \delta', I', H)$ is another rough finite-state automaton, where, for $q \in Q$ and $\sigma \in S$,

$$I' = \bigcup_{q \in I} [q]_R = \bigcup_{q \in I} [q]_{R'}$$

$$\underline{\delta'(q, \sigma)} = \bigcup_{q' \in \delta(q, \sigma)} [q']_R = \bigcup_{q' \in \delta(q, \sigma)} [q']_{R'}$$

and

$$\overline{\delta'(q, \sigma)} = \bigcup_{q' \in \delta(q, \sigma)} [q']_R = \bigcup_{q' \in \delta(q, \sigma)} [q']_{R'}.$$

When $R' \supseteq R$, then A/R' computes with *coarser granules*, while when $R' \subseteq R$, A/R' computes with finer granules.

Table B.1: State transition table for Example B.2.1

Q	$\delta(q,0)$	$\delta(q,1)$
q_0	$\big(\varnothing,\{q_1,q_2\}\big)$	$\big(\{q_3,q_4\}\{q_3,q_4\}\big)$
q_1	$\big(\varnothing,\{q_5,q_6\}\big)$	$\big(\varnothing,\{q_7,q_8\}\big)$
q_2	$\big(\varnothing,\{q_7,q_8\}\big)$	$\big(\{q_7,q_8\}\{q_7,q_8\}\big)$

Theorem B.2.2 *Assume that β and β' are the behaviors of A and A/R', respectively. Then,*

(i) when $R' \supseteq R$ this implies that $\underline{\beta} \subseteq \underline{\beta'}$ and $\overline{\beta} \subseteq \overline{\beta'}$; and

(ii) when $R' = R$ this implies that $\beta = \beta'$.

Proof Suppose that $R' \supseteq R$. Then, $I \subseteq I'$ which means that $\underline{\delta^*(I,x)} \subseteq \underline{\delta^{*'}(I',x)}$ and $\overline{\delta^*(I,x)} \subseteq \overline{\delta^{*'}(I',x)}$. This implies that $\underline{\beta} \subseteq \underline{\beta'}$ and $\overline{\beta} \subseteq \overline{\beta'}$.

Assume that $R' = R$. Then, $I = I'$ which means that $\underline{\delta^*(I,x)} = \underline{\delta^{*'}(I',x)}$ and $\overline{\delta^*(I,x)} = \overline{\delta^{*'}(I',x)}$. This implies that $\underline{\beta} \subseteq \underline{\beta'}$ and $\overline{\beta} \subseteq \overline{\beta'}$ and therefore $\beta = \beta'$. $\square$

Example B.2.1 Suppose that one wants to construct a rough finite-state automaton to recognize whether a patient has flu or not. First, she starts by observing that any person that has headache and high fever definitely suffers from flu. Those that have no headache but high fever or headache but no fever may or may not suffer from flu. Finally, those people who have no headache and no fever do not suffer from flu (Table B.1).

Based on these observations, it is clear that flu is characterized by a rough set of symptoms whose lower approximation is

$$\big\{\text{headache, fever}\big\}$$

and whose upper approximation is

$$\big\{\text{headache, fever, no headache, no fever, headache and fever,}$$
$$\text{no headache but fever, headache but no fever}\big\}.$$

Using these data one can construct a rough finite-state automaton able to recognize flu.

The symptoms headache and fever will be denoted by 1 and absence of a headache or fever will be denoted by 0; thus, $S = \{0, 1\}$. Moreover,

$$Q = \big\{q_0, q_1, q_2, q_3, q_4, q_5, q_6, q_7, q_8\big\},$$
$$Q/R = \big\{\{q_0\}, \{q_1, q_2\}, \{q_3, q_4\}, \{q_5, q_6\}, \{q_7, q_8\}\big\},$$
$$I = \big\{q_0\big\},$$
$$H = \big\{q_2, q_4, q_7, q_8\big\}.$$

Also,

$$\delta(q_0, 0) = \delta(q_0, 1) = \delta(q_0, 00) = \delta(q_0, 01) = \delta(q_0, 10) = \varnothing$$

$$\delta(q_0, 11) = \left\{q_7, q_8\right\} = \overline{\delta(q_0, 11)} = \overline{\delta(q_0, 10)} = \overline{\delta(q_0, 01)}$$

$$\overline{\delta(q_0, 1)} = \left\{q_3, q_4\right\}; \ \overline{\delta(q_0, 0)} = \left\{q_1, q_2\right\}; \ \overline{\delta(q_0, 00)} = \left\{q_5, q_6\right\}.$$

Hence, $\underline{\beta} = \left\{11\right\}$ and $\overline{\beta} = \left\{0, 1, 01, 10, 11\right\}$.

B.3   Rough Turing Machines

A rough Turing machine should compute a result by following the general idea of rough sets, that is, it should compute an upper and a lower approximation of the result. Naturally, the result sought should lie somewhere between the computed results. If these two approximations coincide, then the result of computation is the result sought. In order to be able to compute the upper and the lower approximations, a rough Turing machine should be equipped with two tapes and two scanning heads. Their controlling devices should be different, while initially both tapes should have the same data printed on them. The machine halts when action on both tapes stops.

It is a fact that the transition function of a nondeterministic Turing machine returns a subset of $Q \times \Gamma \times \{L, R, N\}$. In addition, the transition function of an ordinary automaton returns a state, the transition function of a nondeterministic automaton returns a set of states, and the transition function of a rough automaton returns a rough set of states. Therefore, one should expect that the transition function of a rough Turing machine returns a rough set.

The following is a first attempt to formally define a rough Turing machine based on the previous remarks:

Definition B.3.1 A rough Turing machine $\mathscr{M}$ is an octuple $(Q, \Sigma, \Gamma, \delta, \sqcup, \triangleright, q_0, H, R)$, where

- R is an equivalence relation on $Q \times \Gamma \times \{L, R, N\}$;

- δ is a function from $\left((Q \setminus H) \times \Gamma\right) \times \left((Q \setminus H) \times \Gamma\right)$ to $\mathbf{A} \times \mathbf{A}$, where

$$\mathbf{A} = \left\{\left(\underline{A}, \overline{A}\right) \,\middle|\, A \subseteq Q \times \Gamma \times \left\{L, R, N\right\}\right\}$$

is the set of all definable sets in the approximation space $\left(Q \times \Gamma \times \{L, R, N\}, R\right)$, such that $\delta(q_i, s_j, q_k, s_l) = (\underline{A}, \overline{A})$ is a rough set that belongs to $\mathbf{A}$; and

- all others are as in the case of an ordinary Turing machine (see Definition 2.1.2 on p. 14).

References

[1] Akl, S.G.: Three Counterexamples to dispel the myth of the universal computer. Parallel Process. Lett. **16**, 381–403 (2006)

[2] Akl, S.G.: Even accelerating machines are not universal. Int. J. Unconventional Comput. **3**, 105–121 (2007)

[3] Aristotle: Organo: Categoriae and De Interpretatione. Kaktos Editions, Athens (1994)

[4] Atanassov, K.T.: Intuitionistic fuzzy sets. Fuzzy Set. Syst. **20**(1), 87–96 (1986)

[5] Awodey, S.: Continuity and logical completeness: An application of sheaf theory and topoi. In: Benthem, J.V., Heinzmann, G., Rebuschi, M., Visser, H. (eds.) The Age of Alternative Logics, pp. 139–149. Springer (2006)

[6] Barr, M.: Fuzzy set theory and topos theory. Can. Math. Bull. **29**, 501–508 (1986)

[7] Basu, S.: Rough finite-state automate. Cybernet. Syst. Int. J. **36**(2), 107–124 (2005)

[8] Baumeister, R.F.: The Cultural Animal: Human Nature, Meaning, and Social Life. Oxford University Press, Oxford (2005)

[9] Bedregal, B.C., Figueira, S. On the computing power of fuzzy Turing machines. Fuzzy Set. Syst. **159**, 1072–1083 (2008)

[10] Bell, J.L.: Abstract and variable sets in category theory. In: Sica, G. (ed.) What is Category Theory?, pp. 9–16. Polimetrica Publisher, Monza (2006)

[11] Benâtre, J.-P., Métayer, D.L.: The gamma model and its discipline of programming. Sci. Comput. Program. **15**, 55–77 (1990)

[12] Benâtre, J.-P., Métayer, D.L.: Programming by multiset transformation. Commun. ACM **36**(1), 98–111 (1993)

[13] Benâtre, J.-P., Fradet, P., Métayer, D.L.: Gamma and the chemical reaction model: Fifteen years after. In: Calude, C.S., Păun, G., Rozenberg, G., Salomaa, A. (eds.) Lecture Notes in Computer Science, vol. 2235, pp. 17–44. Springer, Heidelberg (2001)

[14] Berry, G., Boudol, G.: The chemical abstract machine. Theor. Comput. Sci. **96**, 217–248 (1992)

A. Syropoulos, *Theory of Fuzzy Computation*, IFSR International Series on Systems Science and Engineering 31, DOI 10.1007/978-1-4614-8379-3,
© Springer Science+Business Media New York 2014

[15] Biacino, L., Gerla, G.: Decidability, recursive enumerability and kleene hierarchy for *L*-subsets. Zeitschrift für mathematische Logik und Grundlagen der Mathematik **34**, 49–62 (1989)

[16] Biacino, L., Gerla, G.: Fuzzy logic, continuity and effectiveness. Archive Math. Logic **41**(7), 643–667 (2002)

[17] Black, M.: Vagueness: An exercise in logical analysis. Philos. Sci. **4**(4), 427–455 (1937)

[18] Boolos, G.S., Burgess, J.P., Jeffrey, R.C.: Computability and Logic, 4th edn. Cambridge University Press, Cambridge (2002)

[19] Byrne, R.M.J., Johnson-Laird, P.N.: 'If' and the problems of conditional reasoning. Trends Cogn. Sci. **13**(7), 282–287 (2009)

[20] Calude, C.S., Păun, G., Rozenberg, G., Salomaa, A. (eds.): Lecture Notes in Computer Science, vol. 2235. Springer, Heidelberg (2001)

[21] Castro, J.L., Delgado, M., Mantas, C.J.: A new approach for the execution and adjustment of a fuzzy algorithm. Fuzzy Set. Syst. **121**, 491–503 (2001)

[22] Cerofolini, G., Amato, P.: Fuzzy chemistry — An axiomatic theory for general chemistry. In: IEEE International Fuzzy Systems Conference 2007 (FUZZ-IEEE 2007), London, UK (2007)

[23] Chang, S.-K.: On the execution of fuzzy programs using finite-state machines. IEEE Trans. Comput. **C-21**(3), 241–253 (1972)

[24] Cheeseman, P.: Discussion: Fuzzy thinking. Technometrics **37**(3), 282–283 (1995)

[25] Church, A.: A set of postulates for the foundation of logic. Ann. Math. **33**(2), 346–366 (1932)

[26] Church, A.: A set of postulates for the foundation of logic (second paper). Ann. Math. **34**(4), 839–864 (1933)

[27] Church, A.: A note on the entscheidungsproblem. J. Symbolic Logic **1**(1), 40–41 (1936)

[28] Church, A.: An unsolvable problem of elementary number theory. J. Math. **58**(2), 345–363 (1936)

[29] Ćirić, M., Stamenković, A., Ignjatović, J.: Factorization of fuzzy automata. In: Csuhaj-Varjú, E., Ésik, Z., (eds.) Proceedings of the Fundamentals of Computation Theory 16th International Symposium, FCT 2007, Budapest, Hungary, August 2007. Lecture Notes in Computer Science, vol. 4639, pp. 213–225. Springer, Berlin (2007)

[30] Clares, B.: Una introduccion a la W-calculabilidad: Operaciones basicas. Stochastica **7**(2), 111–135 (1983)

[31] Clares, B., Delgado, M.: Introduction to the concept of recursiveness of fuzzy functions. Fuzzy Set. Syst. **21**(3), 301–310 (1987)

[32] Cleland, C.E.: On effective procedures. Mind. Mach. **12**, 159–179 (2002)

[33] Conway, J.H., Kochen, S.: The free will theorem. Found. Phys. **36**(10), 1441–1473 (2006)

[34] Conway, J.H., Kochen, S.: The Strong Free Will Theorem. Notice. AMS **56**(2), 226–232 (2009)

[35] Cook, D.B.: Probability and Schrödinger's Mechanics. World Scientific, Singapore (2002)

[36] Cook, S.A., Aanderaa, S.O.: On the minimum computation time of functions. Trans. Am. Math. Soc. **142**, 291–314 (1969)

[37] Copeland, B.J.: The Church-Turing thesis. In: Zalta, E.N. (ed.) The Stanford Encyclopedia of Philosophy, Fall 2002. http://plato.stanford.edu/archives/fall2002/entries/church-turing/

[38] Cotogno, P.: Hypercomputation and the physical Church-Turing thesis. Brit. J. Philos. Sci. **54**(2), 181–223 (2003)

[39] Davis, M.: Computability and Unsolvability. Dover Publications, Inc., New York (1982)

[40] del Castillo-Mussot, M., Dias, R.C.: Fuzzy sets and physics. Rev. Mex. Fís. **39**(2), 295–303 (1993)

[41] Derbyshire, J.: Prime Obsession: Bernhard Riemann and the Greatest Unsolved Problem in Mathematics. Penguin Group, New York (2004)

[42] Deutsch, D.: Quantum theory, the Church-Turing principle and the universal quantum computer. Proc. Roy. Soc. London A **400**, 97–115 (1985)

[43] Dubois, D., Prade, H.: Rough fuzzy sets and fuzzy rough sets. Int. J. Gen. Syst. **17**(2–3), 191–209 (1990)

[44] Düntsch, I.: A logic for rough sets. Theor. Comput. Sci. **179**(1–2), 427–436 (1997)

[45] D'Errico, L., Loreti, M.: Modeling fuzzy behaviours in concurrent systems. In: Laura, L., Italiano, G., Moggi, E. (eds.) Proceedings of the 10th Italian Conference on Theoretical Computer Science, ICTCS'07, pp. 94–105 (2007)

[46] D'Errico, L., Loreti, M.: A Process algebra approach to fuzzy reasoning. In: Carvalho, J.P., Dubois, D., Sousa, J.M.C. (eds.) Proceedings of the Joint 2009 International Fuzzy Systems Association World Congress and 2009 European Society of Fuzzy Logic and Technology Conference, pp. 1136–1141, Lisbon, Portugal, U.K. (2009). ISBN: 978-989-95079-6-8

[47] Eilenberg, S.: Automata, Languages, and Machines, vol. A. Academic, New York (1974)

[48] Etesi, G., Németi, I.: Non-Turing computations via malament-hogarth space-times. Int. J. Theor. Phys. **41**(2), 341–370 (2002)

[49] Gerla, G.: Fuzzy ogic: Mathematical Tools for Approximate Reasoning. Kluwer Academic Publishers, Dordrecht (2001)

[50] Gerla, G.: Effectiveness and multivalued logics. J. Symbolic Logic **71**(1), 137–162 (2006)

[51] Gerla, G.: Multi-valued logics, effectiveness and domains. In: Cooper, S.B., Löwe, B., Sorbi, A. (eds.) Computation and Logic in the Real World, Proceedings of Third Conference on Computability in Europe, CiE 2007, Siena, Italy, June 2007. Lecture Notes in Computer Science, vol. 4497, pp. 336–347. Springer, Berlin (2007)

[52] Girard, J.-Y.: Locus Solum: From the rules of logic to the logic of rules. Math. Struct. Comput. Sci. **11**, 301–506 (2001)

[53] Gödel, K.: On Formally Undecidable Propositions of Principia Mathematica and Related Systems. Dover Publications, New York (1992) (Translated by B. Meltzer; introduction by R.B. Braithwaite)

[54] Goguen, J.: L-fuzzy sets. J. Math. Anal. Appl. **18**, 145–174 (1967)

[55] Goldblatt, R.: Topoi: The Categorial Analysis of Logic. Dover Publications, Mineola (2006)

[56] Goldstine, H.H.: The Computer: From Pascal to von Neumann. Princeton University Press, Princeton (1992)

[57] Gottwald, S.: Many-valued logic and fuzzy set theory. In: Höhle, U., Rodabaugh, S.E. (eds.) Mathematics of Fuzzy Sets: Logic, Topology, and Measure Theory, pp. 5–89. Kluwer Academic Publishers, Dordrecht (1999)

[58] Gottwald, S.: Universes of fuzzy sets and axiomatizations of fuzzy set theory. Part I: Model-based and axiomatic approaches. Stud. Logica **82**, 211–244 (2006)

[59] Gottwald, S.: Universes of fuzzy sets and axiomatizations of fuzzy set theory. Part II: Category theoretic approaches. Stud. Logica **84**, 23–50 (2006)

[60] Gottwald, S.: An early approach toward graded identity and graded membership in set theory. Fuzzy Set. Syst. **161**, 2369–2379 (2010)

[61] Grigoriev, D.Y.: Kolmogoroff algorithms are stronger than Turing machines. J. Sov. Math. (now known as Journal of Mathematical Sciences) **14**, 1445–1450 (1980)

[62] Hájek, P.: Metamathematics of Fuzzy Logic. Kluwer Academic Publishers, Dordrecht (1998)

[63] Harkleroad, L.: Fuzzy recursion, RET's and isols. Math. Logic Quart. 30(26–29), 425–436 (1984)

[64] Harnad, S.: Computation is just interpretable symbol manipulation; cognition isn't. Mind. Mach. 4(4), 379–390 (1995)

[65] Hilbert, D.: Mathematical problems. Bull. Ame. Math. Soc. 8(10), 437–479 (1902) [Translated for the Bulletin of the American Mathematical Society, with the author's permission, by Mary Winston Newson. The original appeared in the Göttinger Nachrichten, pp. 253–297 (1900) and in the Archiv der Mathernatik und Physik, 3d series, vol. 1, pp. 44–63 and 213–237 (1901)]

[66] Horowitz, E., Sahni, S.: Fundamentals of Data Structures in Pascal. Computer Science Press, Rockville (1984)

[67] Hromkovič, J.: Theoretical Computer Science: Introduction to Automata, Computability, Complexity, Algorithmics, Randomization, Communication, and Cryptography. Springer, Berlin (2004)

[68] Hyde, D.: Sorites paradox. In: Zalta, E.N. (ed.) The Stanford Encyclopedia of Philosophy, fall 2008 edn. (2008)

[69] Jones, N.D., Gomard, C.K., Sestoff, P.: Partial Evaluation and Automatic Program Generation. Prentice Hall International, Hemel Hempstead (1993)

[70] Kandel, A., Langholz, G. (eds.): Fuzzy Hardware: Architectures And Applications. Kluwer Academic Publishers, Norwell (1998)

[71] Kaplan, D.M.: Regular expressions and the equivalence of programs. J. Comput. Syst. Sci. 3, 361–386 (1969)

[72] Klaua, D.: An early approach toward graded identity and graded membership in set theory. Math. Nachr. 33(5–6), 273–296 (1967)

[73] Kleene, S.C.: General recursive functions of natural numbers. Math. Ann. 112, 727–742 (1936)

[74] Kleene, S.C., Rosser, J.B.: The inconsistency of certain formal logics. Ann. Math. 36(3), 630–636 (1935)

[75] Klir, G.J., Yuan, B.: Fuzzy Sets and Fuzzy Logic : Theory and Applications. Prentice Hall, Upper Saddle River (1995)

[76] Kolmogorov, A.N.: On the notion of algorithm. Uspekhi Mat. Nauk 8, 175–176 (1953) (The paper is part of the Meetings of the Moscow Mathematical Society paper and the English translation used here is [77])

[77] Kolmogorov, A.N.: On the notion of algorithm. In: Kolmogorov, A.N. (ed.) Information Theory a and the Theory of Algorithms, vol. III. Shiryaev, A.N. (ed.) Mathematics and Its Applications (Soviet Series), vol. 27, p. 1. Kluwer Academic Publishers, Dordrecht (2010)

[78] Kolmogorov, A.N., Uspensky, V.A.: On the definition of an algorithm. Uspekhi Mat. Nauk **13**, 3–28 (1958) (English translation in [79])

[79] Kolmogorov, A.N., Uspensky, V.A.: On the definition of an algorithm. Am. Math. Soc. Translat. Ser. 2 **29**, 217–245 (1963)

[80] Kosko, B.: Fuzziness vs. probability. Int. J. Gen. Syst. **17**(2), 211–240 (1990)

[81] Laviolette, M., Seaman, Jr., J.W., Barrett, J.D., Woodall, W.H.: A probabilistic and statistical view of fuzzy methods. Technometrics **37**(3), 249–261 (1995)

[82] Lee, E., Zadeh, L.A.: Note on fuzzy languages. Inf. Sci. **1**, 421–434 (1969)

[83] Lewis, H.R., Papadimitriou, C.H.: Elements of the Theory of Computation, 2nd edn. Pearson Education, Harlow (1998)

[84] Li, Y.: Approximation and universality of fuzzy Turing machines. Sci. China Ser. F Inf. Sci. **51**(10), 1445–1465 (2008)

[85] Li, Y.: Fuzzy Turing machines: Variants and universality. IEEE Trans. Fuzzy Syst. **16**(6), 1491–1502 (2008)

[86] Li, Y.: Lattice-valued fuzzy Turing machines: Computing power, universality and efficiency. Fuzzy Set. Syst. **160**, 3453–3474 (2009)

[87] Li, Y.: Some results of fuzzy Turing machines. In: IEEE Proceedings of the 6th World Congress in Intelligent Control and Automation, pp. 3406–3409 (2006)

[88] Lipschitz, S.: General Topology. Schaum Publishing Co., New York (1965)

[89] Loeb, D.: Sets with a negative number of elements. Adv. Math. **91**, 64–74 (1992)

[90] Mac Lane, S.: Mathematics: Form and Function. Springer, New York (1986)

[91] Milner, R.: Communication and Concurrency. Prentice Hall, Hemel Hempstead (1989)

[92] Milner, R.: Communicating and Mobile Systems: The π-Calculus. Cambridge University Press, Cambridge (1999)

[93] Minsky, M.: Computation: Finite and Infinite Machines. Prentice-Hall, Englewood Cliffs, NJ, USA, 1967.

[94] Moraga, C. Towards a fuzzy computability? Math. Soft Comput. **6**, 163–172 (1999)

[95] Mordeson, J.N., Malik, D.S.: Fuzzy Automata and Languages: Theory and Applications. Chapman and Hall/CRC, Boca Raton (2002)

[96] Nagel, E., Newman, J.R.: Gödel's Proof. New York University Press, New York (2001) (Edited and with a new forward by Douglas R. Hofstadter)

[97] Ord, T., Kieu, T.D:. The diagonal method and hypercomputation. Brit. J. Philos. Sci. **56**(1), 147–156 (2005)

[98] Papadopoulos, B.K., Syropoulos, A.: Fuzzy sets and fuzzy relational structures as chu spaces. Int. J. Uncertain. Fuzz. Knowledge-Based Syst. **8**(4), 471–479 (2000)

[99] Papadopoulos, B.K., Syropoulos, A.: Categorical relationships between Goguen sets and "two-sided" categorical models of linear logic. Fuzzy Set. Syst. **149**, 501–508 (2005)

[100] Păun, G.: Computing with membranes. J. Comput. Syst. Sci. **61**(1), 108–143 (2000)

[101] Păun, G.: Membrane Computing: An Introduction. Springer, Berlin (2002)

[102] Păun, G., Rozenberg, G., Salomaa, A.: DNA Computing. Springer, Berlin (1998)

[103] Pawlak, Z.: Rough sets. Int. J. Comput. Inf. Sci. **11**(5), 341–356 (1982)

[104] Pawlak, Z.: Vagueness — A rough set view. In: Mycielski, J., Rozenberg, G., Salomaa, A. (eds.) Structures in Logic and Computer Science: A Selection of Essays in Honor of A. Ehrenfeucht, vol. 1261. Lecture Notes in Computer Science, pp. 106–117. Springer, Berlin (1997)

[105] Petzold, C.: The Annotated Turing: A Guided Tour Through Alan Turing's Historic Paper on Computability and the Turing Machine. Wiley Publishing, Inc., Indianapolis (2008)

[106] Pontryagin, L.S.: Foundations of Combinatorial Topology. Graylock Press, Rochester (1952) (Translated from the first (1947) Russian edition by F. Bagemihl, H. Komm, and W. Seidel)

[107] Pykacz, J., D'Hooghe, B., Zapatrin, R.R.: Quantum computers as fuzzy computers. In: Reusch, B. (ed.) Fuzzy Days 2001. Lecture Notes in Computer Science, vol. 2206, pp. 526—535. Springer, Berlin (2001)

[108] Rocha, A., Rebello, M., Miura, K.: Toward a theory of molecular computing. Inf. Sci. **106**, 123–157 (1998)

[109] Rogers, Jr., H.: Theory of Recursive Functions and Effective Computability. The MIT Press, Cambridge (1987).

[110] Russell, B.: Vagueness. Austral. J. Philos. **1**(2), 84–92 (1923)

[111] Santos, E.S.: Maximin sequential-like machines and chains. Theory Comput. Syst. **3**(4), 300–309 (1969)

[112] Santos, E.S.: Probabilistic Turing machines and computability. Proc. Am. Math. Soc. **22**(3), 704–710 (1969)

[113] Santos, E.S.: Fuzzy algorithms. Inf. Contr. **17**, 326–339 (1970)

[114] Santos, E.S.: Computability by probabilistic Turing machines. Trans. Am. Math. Soc. **159**, 165–184 (1971)

[115] Santos, E.S.: Machines, programs and languages. Cybernet. Syst. **4**(1), 71–86 (1974)

[116] Santos, E.S.: Fuzzy and probabilistic programs. Inf. Sci. **10**, 331–345 (1976)

[117] Scott, D.: Some definitional suggestions for automata theory. J. Comput. Syst. Sci. **1**(2), 187–212 (1967)

[118] Searle, J.R.: Mind: A Brief Introduction. Oxford University Press, Oxford (2004)

[119] Silberschatz, A., Galvin, P.B., Gagne, G.: Operating System Concepts. Wiley Inc., New York (2005)

[120] Smyth, M.: Effective given domains. Theor. Comput. Sci. **5**, 257–274 (1977)

[121] Sorensen, R.: Vagueness. In: Zalta, E.N. (ed.) The Stanford Encyclopedia of Philosophy, fall 2008 edn. (2008)

[122] Stamenković, A., Ćirić, M., Ignjatović, J.: Reduction of fuzzy automata by means of fuzzy quasi-orders. CoRR abs/1102.5451 (2011)

[123] Syropoulos, A.: Mathematics of multisets. In: Calude, C.S., Păun, G., Rozenberg, G., Salomaa, A. (eds.) Lecture Notes in Computer Science, vol. 2235, pp. 347–358. Springer, Heidelberg (2001)

[124] Syropoulos, A.: Fuzzifying P systems. Comput. J. **49**(5), 619–628 (2006)

[125] Syropoulos, A.: Yet another fuzzy model for linear logic. Int. J. Uncertain. Fuzz. Knowledge-Based Syst. **14**(1), 131–135 (2006)

[126] Syropoulos, A.: Hypercomputation: Computing Beyond the Church-Turing Barrier. Springer New York, Inc., Secaucus (2008)

[127] Syropoulos, A.: Fuzzy chemical abstract machines. CoRR abs/0903.3513 (2009)

[128] Syropoulos, A.: On nonsymmetric multi-fuzzy sets. Crit. Rev. **IV**, 35–41 (2010)

[129] Syropoulos, A.: Intuitionistic fuzzy P systems. Crit. Rev. **V**, 1–4 (2011)

[130] Syropoulos, A.: On generalized fuzzy multisets and their use in computation. Iranian J. Fuzzy Syst. **9**(2), 115–127 (2012)

[131] Tanaka, K., Mizumoto, M.: Fuzzy programs and their executions. In: Zadeh, L.A., Fu, K.-S., Tanaka, K., Shimura, M. (eds.) Fuzzy Sets and Their Applications to Cognitive and Decision Processes, pp. 41–76. Academic Press, New York (1975)

[132] Taylor, J.R.: An Introduction to Error Analysis: The Study of Uncertainties in Physical Measurements, 2nd edn. University Science Books, Sausalito (1997)

[133] Thiele, R.: Hilbert's twenty-fourth problem. Am. Math. Mon. **110**(1), 1–24 (2003)

[134] Tremblay, J.-P., Sorenson, P.G.: The Theory and Practice of Compiler Writing, p. 16. McGraw-Hill, Singapore (1985)

[135] Turing, A.M.: On Computable Numbers, with an application to the Entscheidungsproblem. Proc. London Math. Soc. **42**, 230–265 (1936)

[136] Tzouvaras, A.: The linear logic of multisets. Logic J. IGPL **6**(6), 901–916 (1998)

[137] Vickers, S.: Topology via logic. In: Cambridge Tracts in Theoretical Computer Science, vol. 6. Cambridge University Press, Cambridge (1990)

[138] Vučković, V.: Basic theorems on Turing algorithms. Publ. l'Institut Math. (Beograd) **1**(15), 31–65 (1961)

[139] Vuillemin, J.: Necessity or Contingency: The Master Argument. Center for the Study of Language and Information, Stanford (1996)

[140] Weihrauch, K.: Computable Analysis: An Introduction. Springer, Berlin (2000)

[141] Wiedermann, J.: Fuzzy Turing machines revised. Comput. Inform. **21**(3), 1–13 (2002)

[142] Wiedermann, J.: Characterizing the super-Turing computing power and efficiency of classical fuzzy Turing machines. Theor. Comput. Sci. **317**, 61–69 (2004)

[143] Yager, R.R.: On the theory of bags. Int. J. Gen. Syst. **13**, 23–37 (1986)

[144] Zadeh, L.A.: Fuzzy sets. Inf. Contr. **8**, 338–353 (1965)

[145] Zadeh, L.A.: Fuzzy algorithms. Inf. Contr. **12**, 94–102 (1968)

[146] Zadeh, L.A.: The concept of a linguistic variable and its application to approximate reasoning—I. Inf. Sci. **8**, 199–249 (1975)

[147] Zadeh, L.A.: Fuzzy sets as a basis for a theory of possibility. Fuzzy Set. Syst. **1**, 3–28 (1978)

[148] Zadeh, L.A.: Discussion: Probability theory and fuzzy logic are complementary rather than competitive. Technometrics **37**(3), 271–276 (1995)

[149] Zadeh, L.A.: Fuzzy logic - Computing with words. IEEE Trans. Fuzzy Syst. **4**, 103–111 (1996)

[150] Zadeh, L.A.: From computing with numbers to computing with words—From manipulation of measurements to manipulation of perceptions. Int. J. Appl. Math. Comput. Sci. **12**, 307–324 (2002)

[151] Zadeh, L.A.: Is there a need for fuzzy logic? Inf. Sci. **178**, 2751–2779 (2008)

[152] Zadeh, L.A.: Computing with words: Principal concepts and ideas. In: Studies in Fuzziness and Soft Computing, vol. 277. Springer, Berlin (2012)

[153] Zheng, X., Weihrauch, K.: The arithmetical hierarchy of real numbers. Math. Logic Quart. **47**(1), 51–65 (2001)

Subject Index

Name Index

A. Syropoulos, *Theory of Fuzzy Computation*, IFSR International Series on Systems Science
and Engineering 31, DOI 10.1007/978-1-4614-8379-3,
© Springer Science+Business Media New York 2014

Printed in the United States
By Bookmasters